GCSE 9-1 geography AQA

Exam Practice

Series editor
Bob Digby

Nicholas Rowles

Sanchez

OXFORD
UNIVERSITY PRESS

Great Clarendon Street, Oxford, OX2 6DP, United Kingdom

Oxford University Press is a department of the University of Oxford. It furthers the University's objective of excellence in research, scholarship, and education by publishing worldwide. Oxford is a registered trade mark of Oxford University Press in the UK and in certain other countries

First published in 2018

British Library Cataloguing in Publication Data

Data available

ISBN 978-019-842348-5

Kindle edition ISBN 978-019-842349-2

10 9 8 7 6 5 4

Printed in Italy by L.E.G.O. SpA

Acknowledgements

We are grateful for permission to reprint the following copyright material:

Cover: watchara/Shutterstock; **p13:** Chris Warham/Alamy Stock Photo; **p20:** Gerry Ellis/Minden Pictures/Getty Images; **p35:** Bob Digby; **p48:** Mark Green/ Alamy Stock Photo; **p66:** Nick Rowles; **p68(tl):** Valerijs Novickis/Shutterstock; **p68(tr):** Dirk Ercken/Shutterstock; **p68(bl):** buddhawut/Shutterstock; **p68(br):** seubsai/Shutterstock; **p71:** Dorset Media Service/Alamy Stock Photo; **p74:** wonganan/iStockphoto; **p77:** Dave Ellison/Alamy Stock Photo; **p86(l):** rvimages/iStockphoto; **p86(r):** alexsl/iStockphoto; **p88:** Courtesy of National Grid; **p104(t):** robertharding/Alamy Stock Photo; **p104(b):** Jo-Anne Albertsen/ Alamy Stock Photo; **p112** (clockwise from top left): ssuaphotos/Shutterstock; M. Shcherbyna/Shutterstock; pajtica/Shutterstock; Creative Nature Media/ Shutterstock; Janelle Lugge/Shutterstock; trialhuni/Shutterstock; Kletr/ Shutterstock; Catmando/Shutterstock; **p113:** Anton Foltin/Shutterstock; **p116:** Missing35mm/iStockphoto; **p121(tl):** Alan Curtis/Alamy Stock Photo; **p121(tr):** Eye Ubiquitous/Alamy Stock Photo; **p121(bl):** Vaughan Sam/Shutterstock; **p121(br):** Background All/Shutterstock; **p123(tl):** Martyn Osman/Alamy Stock Photo; **p123(tr):** robertharding/Alamy Stock Photo; **p123(bl):** Tomas Pavelka/Shutterstock; **p123(br):** aminkorea/Shutterstock; **p125(tl):** Mikadun/ Shutterstock; **p125(tr):** Deborah Vernon/Alamy Stock Photo; **p125(bl):** Gordon Shoosmith/Alamy Stock Photo; **p125(br):** Alex Thomson/Alamy Stock Photo; **p134:** Bob Digby; **p138:** Steven May/Alamy Stock Photo; **p139, 141, 143:** worldmapper.org; **p161(t):** Berndt Fischer/Getty Images; **p161(b):** Aleksandr Kazakevich/Shutterstock; **p162(tl):** sydeen/Shutterstock; **p162(tr):** SUWIT NGAOKAEW/Shutterstock; **p162(bl):** Peter Dazeley/Getty Images; **p162(br):** adirekjob/Shutterstock; **p169:** Prometheus72/Shutterstock; **p172:** Picade LLC/ Alamy Stock Photo; **p176:** YOSHIKAZU TSUNO/Getty Images.

Artwork by Aptara Inc.

Every effort has been made to contact copyright holders of material reproduced in this book. Any omissions will be rectified in subsequent printings if notice is given to the publisher.

Guided answers and mark schemes are available on the Oxford Secondary Geography website: **www.oxfordsecondary.co.uk/aqa_gcse_geog**

Please note: The Practice Paper exam-style questions and mark schemes have not been written or approved by AQA. The answers and commentaries provided represent one interpretation only and other solutions may be appropriate.

Introduction

How to be successful in your exams

If you want to be successful in your exams, then you need to know how you will be examined, what kinds of questions you will come up against in the exam, how to use what you know, and what you will get marks for. That's where this book can help.

How to use this book

This book contains the following features to help you prepare for exams for the AQA GCSE 9–1 Geography specification. It is written to work alongside three other OUP publications to support your learning:

- GCSE 9–1 Geography AQA Student Book
- GCSE 9–1 Geography AQA Revision Guide
- GCSE 9–1 Geography AQA Fieldwork.

An introduction (pages 4–12)

This contains details about:

- the exam papers you'll be taking
- what you need to revise for each exam paper
- how exam papers are marked and how to aim for the highest grades.

On your marks (pages 13–60)

This contains detailed guidance about how to answer questions about specific topics using extended writing for 4, 6 and 9 marks in Papers 1 and 2.

This section contains space for you to write and assess exam answers so that you get to know how to write good quality, focused responses.

Exam papers (pages 61–176)

This section contains two sets of exam papers. These are written to match the style of those you'll meet in the AQA GCSE Geography exam. Each set contains:

a) Exam Papers 1 and 2 which assess your knowledge and understanding of the course
b) Exam Paper 3 which assesses two things.
   - An issues analysis. This assesses your skill in interpreting a pre-release booklet. In the real exam, this will have been issued to you several weeks beforehand. This gives you a chance to understand the geographical issues on which the exam will be based.
   - Fieldwork. You'll have two days fieldwork to draw upon for this part of the exam – one day physical geography and one human.

Each of the exam papers has space for you to write answers, just like a real exam. Your teacher will be able to mark it using a mark scheme which is provided online.

AQA's GCSE Geography specification consists of three units. Each unit contains topics. Each unit is assessed by an exam paper (numbered Paper 1, 2, etc.) with sections for different topics, as follows.

Unit 1 Living with the physical environment

This is assessed by Paper 1 in the examination. It consists of three sections, each containing different physical topics.

- **Section A:** *The challenge of natural hazards* includes topics on Tectonic hazards, Weather hazards, and Climate change. You have to know **all** of these topics.
- **Section B:** *The living world* includes topics on Ecosystems, Tropical rainforests, Hot deserts, and Cold environments. You have to know Ecosystems, Tropical rainforests and **one** of Hot deserts and Cold environments.
- **Section C:** *Physical landscapes in the UK* includes topics on Coastal landscapes, River landscapes, and Glacial landscapes. You have to know about **any two** of these three topics.

In addition, you will have learnt geographical skills (e.g. how to interpret statistics, maps, diagrams or photos). You'll have questions on these in each of the different topics.

Unit 2 Challenges in the human environment

This is assessed by Paper 2 in the examination. It consists of three sections, each containing different human topics.

- **Section A:** *Urban issues and challenges* includes topics on a case study of a major city in **either** a low-income country (LIC) **or** a newly emerging economy (NEE), **and** a case study of a major city in the UK.
- **Section B:** *The changing economic world* includes topics on a case study of **either** an LIC **or** an NEE, **and** Economic futures in the UK.
- **Section C:** *The challenge of resource management* includes topics on Resource management and Food, Water and Energy. You have to know about Resource management and **one** of Food **or** Water **or** Energy.

In addition, you will have learnt geographical skills (e.g. how to interpret statistics, maps, diagrams or photos). You'll have questions on these in each of the different topics.

Unit 3 Geographical applications

This is assessed by Paper 3 in the examination. It consists of two sections.

- **Section A:** *Issue evaluation*. You don't need to learn any topics for this, but you must know and understand the pre-release booklet thoroughly.
- **Section B:** *Fieldwork*. You need to revise your two fieldwork topics and apply what you learnt in those topics.

Of the three exam papers, Papers 1 and 2 are similar, while Paper 3 is very different.

Format of Paper 1

- **Time:** 1 hour 30 minutes.
- **Worth:** 88 marks in total – 85 on topics you've learnt, and another 3 for spelling, punctuation, grammar and the use of specialist geographical terminology, which is assessed on one question in Section A.
- **Counts for:** 35% of your final grade.
- **Number of sections:** three, assessing the topics described in Unit 1 on page 5.

You must answer:

- **all** parts of Section A – with questions on Tectonic hazards, Weather hazards, and Climate change. This section has 33 marks.
- **all** parts of Section B – with questions on Ecosystems, Tropical rainforests, Hot deserts **or** Cold environments. This section has 25 marks.
- **any two** questions in Section C, chosen from Question 3 (River landscapes in the UK), Question 4 (Coastal landscapes in the UK), or Question 5 (Glacial landscapes in the UK). This section has 30 marks (i.e. 15 marks each).

Any resources that you need to answer the questions are included – there is no separate resource booklet.

Format of Paper 2

- **Time:** 1 hour 30 minutes.
- **Worth:** 88 marks in total – 85 on topics you've learnt, and another 3 for spelling, punctuation, grammar and the use of specialist geographical terminology which is assessed on one question in Section A.
- **Counts for:** 35% of your final grade.
- **Number of sections:** three, assessing the topics described in Unit 2 on page 5.

You must answer

- **all** parts of Section A – with questions on a case study of a major city in **either** an LIC **or** an NEE **and** a case study of a UK major city. This section has 33 marks.
- **all** parts of Section B – with questions on a case study of **either** an LIC **or** an NEE **and** Economic futures in the UK. This section has 30 marks.
- **any one** question in Section C, chosen from Question 3 (Food), Question 4 (Water), or Question 5 (Energy). This section has 25 marks.

Any resources that you need to answer the questions are included – there is no separate resource booklet.

Format of Paper 3

- **Time:** 1 hour 15 minutes.
- **Worth:** 76 marks in total – 70 for geographical questions, and two lots of 3 marks for spelling, punctuation, grammar and the use of specialist geographical terminology which are each assessed in Sections A and B.
- **Counts for:** 30% of your final grade.
- **Number of sections:** two, assessing Issue evaluation and Fieldwork.

You must answer

- **all** of Section A, with questions on the Issue evaluation. This section has 37 marks.
- **all** of Section B with questions about your two days' fieldwork (one physical, one human, with some understanding of the links between them). This section has 39 marks.

Section A uses a separate resource booklet. You will already have seen a copy of this, and have worked on it in class. However, you won't be allowed to take that copy in – you'll be issued with a new copy. Section B includes resources which will appear alongside the questions.

Question style

In each section, the first questions are short, each worth 1 or 2 marks.

- These include a mix of multiple-choice, short answers or calculations.
- There are resource materials (data, photos, cartoons, etc.) on which you'll be asked questions. These could include stats skills, so remember you can use a calculator in each exam.
- You'll be expected to know what these resources are getting at from what you have learnt.
- Detailed case study knowledge is NOT needed on shorter questions like these – though you can get marks for using examples.

All these questions are **point-marked** (see page 9).

Later questions on each topic require extended writing, and are worth 4, 6 or 9 marks. You need to have learnt examples and case studies to answer these questions. Answers like this are marked using **levels** – from Level 1 (lowest) to Level 3 (highest) – see pages 10 and 11.

These questions could be based on any part of a topic, but are likely to be on those parts where you have been taught examples (e.g. a tectonic hazard, or a UK extreme weather event).

One 9-mark question on each paper is assessed for spelling, punctuation, grammar and use of specialist geographical terminology for 3 marks – making it worth 12 marks in total.

Answering questions properly is the key to success. When you first read an exam question, check out the **command word** – that is, the word that the examiner uses to tell you what to do. The table below gives you the command words you can expect, and the number of marks you can expect for each command word.

Command words

Command word	Typical no. of marks	What the command word means	Example of a question
Identify/State/ Name	1	Find (e.g. on a photo), or give a simple word or statement	Identify the landform in the photo
Define	1	Give a clear meaning	Define the term 'fertility rate'
Calculate	1 or 2	Work out	Calculate the mean depth of the river shown in Figure 2
Label	1 or 2	Print the name of, or write, on a map or diagram	Label two features of the cliff in Figure 4
Draw	1, 2 or 3	As in sketch or draw a line	Draw a line to complete the graph in Figure 3
Compare	2 or 4	Identify similarities or differences	(*referring to a graph*) Compare the rate of population growth in city X with city Y.
Describe	2 or 4	Say what something is like; identify trends (e.g. on a graph)	Describe the trend shown in Figure 1
Explain	2, 4, 6 or even 9	Give reasons why something happens	Using examples, explain the rapid growth of a mega-city you have studied
Suggest	2 or 4	In an unfamiliar situation (e.g. a photo or graph), explain how or why something might occur, with a reason	Suggest reasons for the increase shown in the graph
Examine	6 or 9	Give reasons for, but also begin to judge which of the reasons is more important	Examine the reasons for the growth of one mega-city you have studied
To what extent …	6 or 9	Show how far you agree or disagree with a statement	To what extent do mega-cities offer a better lifestyle for migrants than the rural areas they have left?
Assess	6 or 9	Weigh up which is most/least important	Assess the need for coastal management along a stretch of coast you have studied
Evaluate	6 or 9	Make judgements about which is most or least effective	Evaluate the methods used in collecting data in your fieldwork
Discuss	6 or 9	Give an overview of a situation or a topic where there are different approaches or viewpoints	Discuss the ways in which climate change could be managed
Justify	6 or 9	Give reasons why you support a particular decision or opinion	Justify the reasons for your choice

Examiners have clear guidance about how to mark. They must mark fairly, so that the first candidate's exam paper in a pile is marked in exactly the same way as the last.

You will be rewarded for what you know and can do; you won't lose marks for what you leave out. If your answer matches the best qualities in the mark scheme then you'll get full marks.

Any questions which carry 1 or 2 marks are **point marked**, and those which carry 4 marks or more are **level marked**. Be clear about what this means.

Point marking

Look at this question.

> Rainforests are ecosystems. State **two** ways in which humans can protect ecosystems.
>
> **[2 marks]**

There are 2 marks for this question and you must therefore suggest two ways of protecting ecosystems. So, you get 1 mark (or point) for each way that you give. The mark scheme tells examiners the ways they can mark an answer.

Extending an answer

The key to success is knowing how to turn 1 mark into 2, or 2 into 3. To achieve this, you need to learn how to **develop** answers. Imagine if the question above had been:

> Rainforests are ecosystems. Describe **one** way in which humans can protect ecosystems.
>
> **[2 marks]**

In this case, it is not enough to name one way of protecting an ecosystem – this would earn just 1 mark. To earn 2 marks your answer must be either:

- **extended** – i.e. describing in more detail, or
- **exemplified** – i.e. giving an example of what you are describing, or
- **explained** – i.e. giving a reason why something occurs.

Now consider this question:

Explain one possible economic impact of climate change.

[2 marks]

This question asks for an explanation about an impact, i.e. not just stating what the impact is but why it occurs. It's the explanation that earns marks. In the examples below, a tick is used to show where marks are earned.

> Sea level change might mean farmland near the coast gets flooded. ✓

This is an impact – but it is only worth 1 mark as it simply says one thing.

> Sea level change might mean farmland near the coast gets flooded ✓ **so that** farmers lose their crops. ✓

However, this has been **extended** to make it worth a second mark. The connecting phrase 'so that' is one of the most useful in helping you get higher marks!

Other useful connecting phrases include 'therefore', or 'which leads to'.

> Sea level change might mean farmland near the coast gets flooded ✓ **because farmers would lose crops e.g. rice.** ✓

This is an economic impact which has been developed with an **example**, so it gains a second mark.

> Sea level change might mean farmers near the coast might lose rice crops ✓ if saltwater flooded the fields **as rice is a freshwater plant**. ✓

This is an example of explanation of why crops are lost, so it gets a second mark.

Level marking

Longer, extended questions which carry 4, 6 and 9 marks are always marked using levelled mark schemes. Examiners mark extended answers based on levelled mark schemes.

There are three levels, with Level 1 the lowest and Level 3 the highest.

- Level 1 answers are basic, written in very general terms, with few or no examples of real places; they contain few geographical words, and writing is often rambling, rather than structured.
- Level 3 answers are detailed, are almost always about real places with data and examples, and are explained in a structured way.

In these cases, it is not the number of points made, but the ways in which points are extended that matters. Look at this question.

Assess the effects of a named tropical storm.

[6 marks]

Below is a mark scheme which uses levels of response. In marking questions, examiners do not mark points, but instead read the answer as a whole, and judge it against the qualities shown in the mark scheme. They are mainly looking for ways in which the answer is developed and explained – about three extended points are needed for maximum marks. Notice that Level 3 is reserved for candidates who 'assess' as the command word tells them to do. Candidates who just explain the effects can only get Level 1 or 2.

Level	Marks	Description
3 (Detailed)	5–6	Explains very clearly. Reaches judgement about the greatest or least effects of the storm, based on evidence. Well organised, and explains effects well, using extended points to develop the answer. Includes a range of specific and located impacts which are detailed. Well written with good use of geographical words. Spelling, punctuation and grammar always accurate.
2 (Clear)	3–4	Begins to explain fairly clearly. Two or three effects are explained briefly and with some extension. Little or no judgement about greatest or least effects of storm. Examples are used, but vary in detail; places (e.g. countries) and impacts are named. Clearly written, with some use of geographical terms. Spelling, punctuation and grammar vary but are generally accurate.
1 (Basic)	1–2	Very limited or no reasoning. One or two effects are simply described but are not extended. Most of the answer is generalised without detail or use of named examples. Places are poorly located (e.g. 'in India'). Few geographical terms or phrases. Spelling, punctuation and grammar are weak.
	0	No accurate response

What makes a good Level 3 answer?

Below is a Level 3 answer to the 6-mark question, 'Assess the economic effects of a named tropical storm.'

The storm is named. Always do this – you could limit yourself to Level 1 or 2 if you don't. The **name** doesn't actually get a mark, but it makes it possible to get the maximum mark.

'In Cyclone Aila which affected Bangladesh in 2009 the effects were enormous, especially among the poor. The cyclone brought extreme high winds and floods. This damaged buildings, which cost insurance companies and governments millions. More storms also caused erosion of flood defences, which flooded villages and farmland costing huge amounts to replace and repair, and destroyed crops. For many farmers and families, this meant loss of homes and crops, making their poverty worse and forcing some to leave the land and move to Dhaka, the capital, for work.'

'High winds and floods' are the first economic effects. Credit is given for the detailed explanation of the economic effects of a storm – ***'damaged buildings', 'cost insurance companies and governments millions'.*** Notice these are linked by the connective phrase ***'which cost'***. This turns a Level 1–2 point about 'damaging buildings' into Level 3 by adding more detail.

'Erosion of flood defences' is also an economic impact. It is extended with ***'flooded villages and farmland'*** and ***'costing huge amounts to replace and repair'***. The phrase ***'costing huge amounts'*** is a judgement of its seriousness, making this answer Level 3.

'loss of homes and crops' is evidence of another economic impact. This is extended by the candidate who explains the seriousness of loss of crops – making a judgement and answering the command word 'Assess'.

Always answer the question that is set, not the one you would like! Good answers are often shorter and more focused – longer ones are more likely to stray off the point. To help you focus, unpick the question, as shown below.

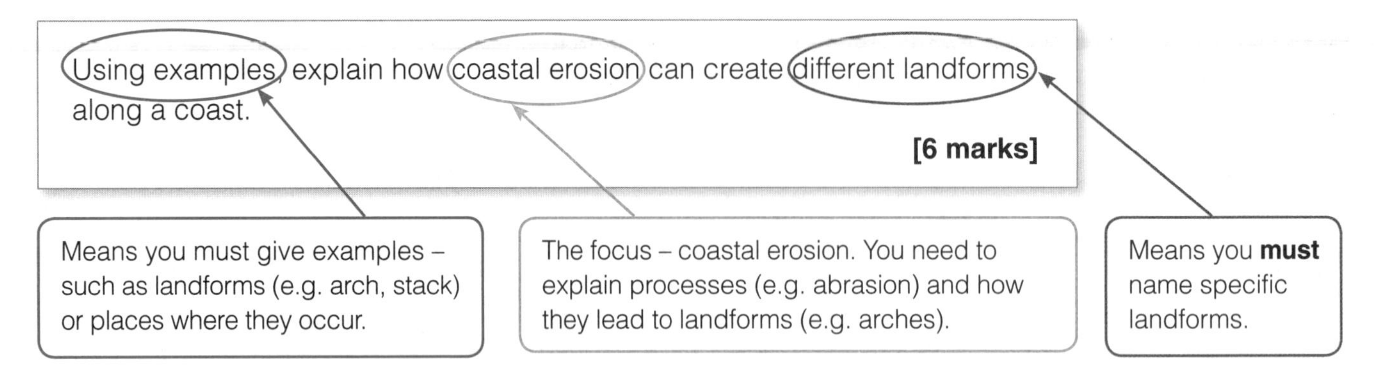

This question is about explaining processes that lead to different landforms (e.g. caves, arches or cliffs). You must explain and give reasons why these processes happen – don't just describe. Good explanation takes you to Level 3.

Using case studies

Candidates often worry about how to write a good case study. The following question requires the use of examples.

> Using examples, assess the success of regeneration in one named UK city.
>
> **[6 marks]**

This question can be answered using the example of Bristol (see Student Book 14.10 and 14.11). Use other examples if you have studied them or carried out urban fieldwork, e.g. in east London. Plan your case studies using a spider diagram to help organise your thoughts, as below.

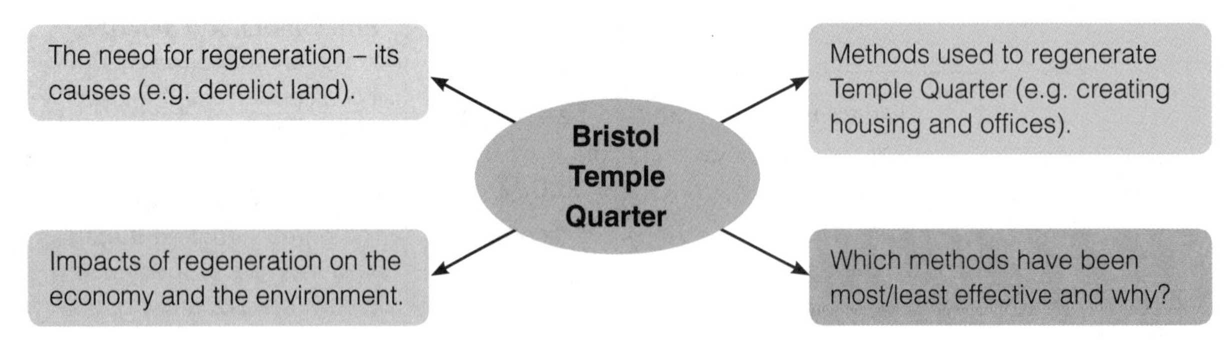

From a spider diagram, you can build up detailed notes, such as:

- The need for regeneration – draw two 'legs' from this box to explain economic causes (e.g. lack of jobs) and environmental (e.g. derelict land).
- Methods used to regenerate Temple Quarter – draw three 'legs' to explain methods used to attract people, build the economy and improve the environment.
- Impacts of regeneration on the economy and the environment.
- Finally, draw 'legs' to show which methods have been effective and why.

On your marks

Nailing the 4-mark questions

- **In this section you'll learn how to maximise marks on 4-mark questions.**

The question below has a 4-mark tariff, which is marked using levels based upon the quality of the answer as a whole, rather than by individual points (see page 15).

Study **Figure 1**. It shows a house damaged by Typhoon Haiyan in a rural area of the Philippines in 2013.

Figure 1

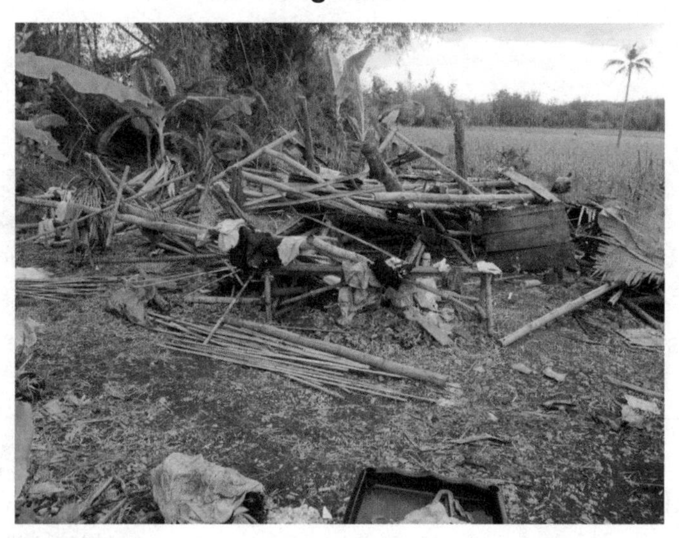

Using **Figure 1** and your own knowledge, suggest two reasons why tropical storms have such a big impact on developing countries like the Philippines.

[4 marks]

1 Plan your answer

Before attempting to answer the question, remember to **BUG** it. That means:

✓ **Box** the command word.
✓ **Underline** the following:
- the theme
- the focus
- any evidence required
- the number of examples needed.

✓ **Glance** back over the question, to make sure you include everything in your answer.

Use the BUG on page 14 to plan your own answer.

Five steps to success!

The following five steps are used in this chapter to help you get the best marks.

1. **Plan your answer** – decide what to include and how to structure your answer.
2. **Write your answer** – use the answer spaces to complete your answer.
3. **Mark your answer** – use the mark scheme to self- or peer-mark your answer. You can also use this to assess sample answers in step 4 below.
4. **Sample answers** – sample answers are given to show you how to maximise marks for a question.
5. **Marked sample answers** – these are the same answers as for step 4, but are marked and annotated, so that you can compare these with your own answers.

Evidence: Support your answer with information from the photo AND from your own knowledge. You must do both to get 4 marks.

Command word: Make intelligent, reasoned interpretations.

Using Figure 1 and your own knowledge, suggest two reasons why tropical storms have such a big impact on developing countries like the Philippines. **[4 marks]**

Theme: This question is linked to the theme of Weather hazards, assessed in Paper 1, Section A of your exam. The question is compulsory.

Focus: You must consider why the hazard has such an impact, not just describe what you see in the photo.

Focus and number of examples: The question asks for two specific reasons about developing countries, so you must offer two AND expand them with more detail.

PEEL your answer

Use PEEL notes to structure your answer. This will help you to communicate your ideas to the examiner in the clearest way. PEEL has four stages.

- **P**oint – make two points for this question. Use sentences, not bullet points.
- **E**xplain – give reasons for each point. Use sentence starters such as: 'This is because ...', 'One reason is ...'.
- **E**vidence – include facts and other details from named examples to back up each point.
- **L**ink – link the two points to each other, use PEE sentence starters such as: 'A second way is ...' or 'Secondly ...'. You'll learn more about how to do this in the sections that follow about 6- and 9-mark questions.

Quality not quantity

You will not be marked simply on the number of points you make, but on the quality of your answer. That means the quality of the content and also how well you structure your answer.

2 Write your answer

Using **Figure 1** and your own knowledge, suggest two reasons why tropical storms have such a big impact on developing countries like the Philippines. **[4 marks]**

1. ______________________________

____

2. ______________________________

____

3 Mark your answer

1. To help you to identify well-structured points in the answer, highlight the:
   - points in red
   - explanations in orange
   - evidence in blue.

2. Use the mark scheme below to decide what mark to give. Four-mark questions are not marked using individual points, but instead you should choose a level and a mark based upon the quality of the answer as a whole.

Level	Marks	Description	Examples
2 (Clear)	3–4	• Shows accurate understanding of impacts by applying relevant knowledge and understanding to the photo. • Makes clear and effective use of the photo to explain the impacts of a tropical storms.	• *Many buildings in developing countries are built of materials which would collapse during a tropical storm, because there are few building regulations.* • *The photo shows poor quality housing which could not withstand the high winds of a tropical storm.*
1 (Basic)	1–2	• Shows limited understanding of impacts by applying some knowledge and understanding to the photo. • Makes limited and piecemeal use of the photo to explain the impacts of a tropical storm.	• *Houses in many poorer countries are built from things like wood which would fall down during a big storm.* • *Many people who live in developing countries do not have the money to build houses with proper materials.*
	0	No accurate response	

4 Sample answer

Read through the sample answer. Go through it using the three colours in Section 3 above, and decide how many marks it is worth.

Question recap

Using **Figure 1** and your own knowledge, suggest two reasons why tropical storms have such a big impact on developing countries like the Philippines.

Sample answer 1

1. The photo shows a wooden building which has fallen down, probably during the high winds in a tropical storm which will have destroyed it. The people who live there are probably poor and have no resistance to storms like this.

2. In the photo, different building materials (bits of wood, bamboo) show the building was cheap to build, but weak. This is typical of developing countries where many people might be very poor.

Strengths of the answer			
Ways to improve the answer			
Level		**Mark**	

 Marked sample answer

Sample answer 1 is marked below. The text has been highlighted as follows to show how well the answer has structured points:

- points in red
- explanations in orange
- evidence in blue.

Marked sample answer 1

Evidence – wooden buildings visible in the photo

Point – the building has fallen down – an essential part of the answer

Explanation – suggests that high winds have destroyed this building and makes the link to tropical storms

1. The photo shows a wooden building which has fallen down, probably during the high winds in a tropical storm which will have destroyed it. The people who live there are probably poor and have no resistance to storms like this.

Explanation – suggests people are poor with no resistance to tropical storms

2. In the photo, different building materials (bits of wood, bamboo) show the building was cheap to build, but weak. This is typical of developing countries where many people might be very poor.

Evidence – specific building materials from the photo

Point – the building was cheap to build – links to part one of the answer above

Explanation – there is no explanation here, but two have been offered above. In a 4-mark question, this is sufficient

 **Examiner feedback**

The full descriptor for Level 2 applies to this answer as follows:

- '*Shows accurate understanding of impacts …*' – sees the link between the high winds of a tropical storm and the quality of the building and links this to what can be seen in the photo.
- '*… by applying relevant knowledge and understanding to the photo*' – understands the weather conditions of a tropical storm, and how this can affect weak buildings; it also links to the building materials and their 'cheapness'.
- '*Makes clear and effective use of the photo to explain the impacts of a tropical storm*' – in both answers, the student refers directly to evidence which is visible.

By meeting the descriptors fully, the answer earns all 4 marks.

Now try this one!

Follow the stages from the previous example to tackle this 4-mark question.

Study **Figure 2**. It shows measures of development for three countries.

Figure 2

Country	HDI	Death rate per 1000 population	% population with access to safe water
Japan	0.891	9.51	100
Brazil	0.755	6.58	98
Zimbabwe	0.509	10.13	77

Explain the strengths and limitations of any **one** of the indicators in **Figure 2** in seeking to understand a country's level of development.

[4 marks]

1 Plan your answer

Before attempting to answer the question, remember to **BUG** it. Use the guidelines on page 13. Annotate it in the margin.

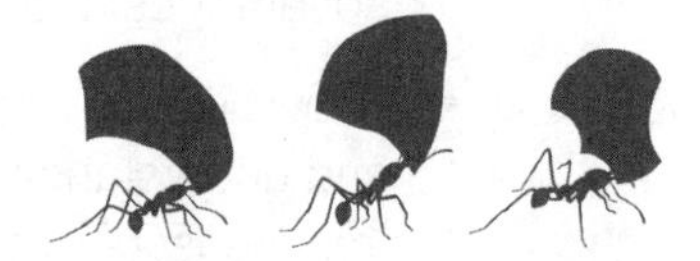

Explain the strengths and limitations of any **one** of the indicators in **Figure 2** in seeking to understand a country's level of development.

[4 marks]

PEEL your answer

Use PEEL notes to structure your answer. Refer to the guidelines on page 14.

2 Write your answer

Explain the strengths and limitations of any **one** of the indicators in **Figure 2** in seeking to understand a country's level of development.

[4 marks]

3 Mark your answer

1. To help you to identify well-structured points in the answer, highlight the:
   - points in red
   - explanations in orange
   - evidence in blue.
2. Use the mark scheme below to decide what mark to give. Four-mark questions are not marked using individual points, but instead you should choose a level and a mark based upon the quality of the answer as a whole.

Level	Marks	Description	Examples
2 (Clear)	3–4	• Shows sound understanding of one measure of development, and its advantages and disadvantages. • Shows sound application of knowledge and understanding in interpreting how that measure of development can show a positive or negative picture.	• *The % of people with access to safe water is useful as it shows water quality results from higher spending on water.* • *Water quality reflects well on a country's level of development since the higher a country's HDI, the better its water, and in turn this would lead to better health.*
1 (Basic)	1–2	• Shows limited understanding of one measure of development, and its advantages and disadvantages. • Shows limited application of knowledge and understanding in interpreting how that measure of development can show a positive or negative picture of a country.	• *Death rates show how good the health of a country is, because people have good health if the figure is low.* • *Death rates are high in developing countries and low in developed countries because health is worse.*
	0	No accurate response	

4 Sample answer

Read this sample answer. Go through it using the three colours in above, and decide how many marks it is worth.

Question recap

Explain the strengths and limitations of any **one** of the indicators in **Figure 2** in seeking to understand a country's level of development.

Sample answer 2

HDI is a good measure of a country's development, because it shows how well developed a country is socially as well as economically. It is a single figure which combines GDP (to show how wealthy a country is) with literacy (which shows the level of education), and life expectancy. So it is a good way of showing how much money is spent on health and education. HDI has a disadvantage because wealthy countries might not have a high HDI figure if wealth is concentrated in the hands of a few wealthy people (like Saudi Arabia), and so does not get spent on the people.

Strengths of the answer			
Ways to improve the answer			
Level		**Mark**	

5 Marked sample answer

Sample answer 2 is marked below. As with marked sample answer 1, the text is highlighted as follows:

- points in red
- explanations in orange
- evidence in blue.

Marked sample answer 2

Point – student immediately makes a judgement which helps to answer the question

Explanation – explains the basic point about HDI by saying what it measures

HDI is a good measure of a country's development, because it shows how well developed a country is socially as well as economically. It is a single figure which combines GDP (to show how wealthy a country is) with literacy (which shows the level of education), and life expectancy. So it is a good way of showing how much money is spent on health and education. HDI has a disadvantage because wealthy countries might not have a high HDI figure if wealth is concentrated in the hands of a few wealthy people (like Saudi Arabia), and so does not get spent on the people.

Evidence – a really clear statement showing the evidence to support the student's judgement about HDI

Point – only a basic point but it helps to balance the answer with a disadvantage as well as advantage

Explanation – an explanation of the point made

Examiner feedback

The full descriptor for Level 2 applies to this answer as follows:

- '*Shows sound understanding of one measure of development …*' – the student clearly knows what HDI is and how it is measured, but also sees the relevance to the question by explaining its advantage in combining three measures together.
- '*…and its advantages and disadvantages*' – makes two points about advantages by explaining how HDI is calculated, and why it does not work in countries such as Saudi Arabia.
- '*Shows sound application of knowledge and understanding in interpreting how that measure of development can show a positive or negative picture of a country*' – in the answer, the student refers directly to evidence which helps to explain the points above.

By meeting the descriptors fully, the answer earns all 4 marks.

On your marks

Using the command word 'Discuss'

- **In this section you'll learn how to tackle 6-mark questions that use 'Discuss' as a command word.**

Six-mark questions – what's different?

By now, you should have got the hang of answering 4-mark questions. Six-mark questions differ from 4-mark questions because they are marked using three levels, not two.

- Levels 1 and 2 are of the same standard as Levels 1 and 2 in the 4-mark questions.
- Level 3 is more challenging and is worth 5–6 marks.

Level 3 means writing to a higher standard.

- Tougher command words are used. 'Explain' or 'Suggest' tend to be used in 4-mark questions. 'Discuss' or 'Assess' tend to be used in 6-mark questions (therefore, two command words in one question).
- Alternatively, you could be given a statement for which you have to provide evidence (see the example below).

A **4-mark question** might ask:

'Explain two impacts of a volcanic eruption on climate'.

A **6-mark question** might ask:

'Volcanic eruptions can cause important changes to the global climate'.
Use evidence to support this statement.

Study **Figure 1**. It shows a rainforest in Borneo, Indonesia which has been cleared to make way for a plantation.

Figure 1

Using **Figure 1** and your own knowledge, discuss the impacts of rainforest clearance.

[6 marks]

Five steps to success!

The following five steps are used in this chapter to help you get the best marks.

1. **Plan your answer** – decide what to include and how to structure your answer.
2. **Write your answer** – use the answer spaces to complete your answer.
3. **Mark your answer** – use the mark scheme to self- or peer-mark your answer. You can also use this to assess sample answers in step 4 below.
4. **Sample answers** – sample answers are given to show you how to maximise marks for a question.
5. **Marked sample answers** – these are the same answers as for step 4, but are marked and annotated, so that you can compare these with your own answers.

1 Plan your answer

Before attempting to answer the question, remember to **BUG** it.
That means:

✓ **Box** the command word.
✓ **Underline** the following:
- the theme
- the focus
- any evidence required
- the number of examples needed.

✓ **Glance** back over the question – to make sure you include everything in your answer.

Use the BUG below to plan your own answer.

Evidence: Support your answer with information from the photo AND from your own knowledge. You must do both to get 6 marks.

Command word: As well as explaining 2 to 3 impacts, you need to decide whether these are positive or negative.

Using Figure 1 and your own knowledge, discuss the impacts of rainforest clearance. **[6 marks]**

Theme: This question is linked to the theme of The Living World, assessed in Paper 1, Section B of your exam. The question is compulsory.

Focus: You must explain reasons why rainforest clearance can have such impacts, based on what you've studied, as well as what's in the photo.

Focus and number of examples: The focus is impacts – i.e. things affected by forest clearance. Six-mark questions don't state how many impacts, but you probably need to explain and develop two, written in paragraphs.

PEEL your answer

Use PEEL notes to structure your answer. This will help you to communicate your ideas to the examiner in the clearest way. PEEL has four stages:

- **P**oint – give three impacts for this question. Use sentences, not bullet points.
- **E**xplain – give reasons for each point. Use sentence starters such as: 'This is because ...', 'One reason is ...'.
- **E**vidence – include details from named examples to support each point.
- **L**ink – link the impacts to each other. Use PEE sentence starters such as: 'A second way is .....' or 'Secondly ...'. Finish it off with a one-sentence conclusion.

Planning grid

Use this planning grid to help you write high-quality paragraphs. Remember to include links to show how your points relate to each other and to the question.

	PEE Paragraph 1	PEE Paragraph 2
Point		
Explanation		
Evidence *(from photo or your own knowledge)*		
Link – a mini conclusion		

2 Write your answer

Using **Figure 1** and your own knowledge, discuss the impacts of rainforest clearance.

[6 marks]

Tip

'Discuss' means use a range of examples!

Don't just describe and explain. If the question asks you to 'Discuss', it wants you to cover a range of impacts.

For example, 'Discuss the impacts of rainforest clearance' would want you to say whether impacts are positive or negative, or perhaps whether they are economic, social or environmental.

One way of doing this is in a **mini-conclusion** – it need only be a sentence or two.

Strengths of the answer			
Ways to improve the answer			
Level		**Mark**	

3 Mark your answer

1. To help you to identify well-structured points in the answer, highlight the:
   - points in red
   - explanations in orange
   - evidence in blue.

2. Use the mark scheme below to decide what mark to give. Six-mark questions are not marked using individual points, but instead you should choose a level and a mark based upon the quality of the answer as a whole.

Level	Marks	Description	Examples
3 (Detailed)	5–6	• Provides a balanced discussion with well-developed impacts and detailed understanding of these impacts, and is able to make a judgement about them. • Shows thorough identification of the evidence for the impacts of forest clearance, and understands the implications.	• *The removal of forest cover would expose the soils to heavy tropical rains, which would erode them, making the land useless for farming.* • *The bare soil in the photo shows how exposed it would be to wind or rain, or tropical sun.*
2 (Clear)	3–4	• Shows accurate understanding of impacts by applying relevant knowledge and understanding to the photo. • Makes clear and effective use of the photo to explain the impacts of clearing the forest.	• *Removing the trees would mean less protection for the soil from heavy rain – runoff would occur and probably take the soil with it.* • *The photo shows little vegetation to protect the soil so it would probably be lost.*
1 (Basic)	1–2	• Shows limited understanding of impacts by applying some knowledge and understanding to the photo. • Makes limited and piecemeal use of the photo to explain the impacts of clearing the forest.	• *The land is all bare and there are no trees there. The rain would wash it away.* • *The photo shows all the trees have been cut and burned and there's nothing there.*
	0	No accurate response	

4 Sample answer

Read these two sample answers 1 and 2 below.

a) Go through each one using the three colours on page 24.

b) Use the level descriptors to decide how many marks it is worth.

Question recap

Using **Figure 1** and your own knowledge, discuss the impacts of rainforest clearance.

Sample answer 1

The forest looks like it has been cleared by burning. The land looks full of tree roots meaning that it will not be easy to plant crops. The soil is black which is probably ash from all the burnt trees after the fires have gone out. The next time it rains, the ash will probably get washed away because it looks like the land is sloping a bit, and farmers might find there is no soil left by the time they get to plant their crops. There is no wildlife, which probably got killed in the fires. Lots of rainforests get cleared by burning like this.

Strengths of the answer			
Ways to improve the answer			
Level		**Mark**	

Sample answer 2

Clearing rainforest areas like this can be disastrous. The forest cover has been cleared and burned. It used to shelter the soil so that when it rained heavily the rain would be intercepted and would drip slowly into the soil. Now if there's a storm, the rain will run off and will probably take the soil with it because there's nothing to protect it.

Soils in rainforests are infertile, so that clearing land for plantations may not be a good idea anyway. Probably the ash that's in the photo would be the only fertile part of the soil. So clearing forests is not likely to have any benefits for farmers.

Strengths of the answer			
Ways to improve the answer			
Level		**Mark**	

5 Marked sample answer

Sample answers 1 and 2 are marked below. The text has been highlighted as follows to show how well each answer has structured points:

- points in red
- explanations in orange
- evidence in blue.
- judgements are underlined. These are important in order to reach Level 3 on questions with the command word 'Discuss'.

Marked sample answer 1

Evidence – evidence from the photo but this is not linked to impacts of clearing rainforest

The forest looks like it has been cleared by burning. The land looks full of tree roots meaning that it will not be easy to plant crops. The soil is black which is probably ash from all the burnt trees after the fires have gone out. The next time it rains, the ash will probably get washed away because it looks like the land is sloping a bit, and farmers might find there is no soil left by the time they get to plant their crops. There is no wildlife which probably got killed in the fires. Lots of rainforests get cleared by burning like this.

Explanation – this brief explanation does tackle one of the impacts of clearance

Evidence – more evidence from the photo but, again, this is not about impacts

Point – makes a point which describes an impact of clearing the forest

Explanation – explains one of the impacts of clearing forest, i.e. soil erosion

Evidence – a third observation from the photo though, again, this is not about impacts

Examiner feedback

The descriptor for Level 2 just about applies to this answer as follows:

- '*Shows accurate understanding of impacts ...*' – the student does understand some of the impacts of rainforest clearance. This is general rather than accurate understanding. For example, there is little subject terminology (e.g. soil erosion).
- '*... by applying relevant knowledge and understanding to the photo*' – the two explanations demonstrate that the student is aware of what rainforest clearance can lead to.
- '*Makes clear and effective use of the photo to explain the impacts of clearing the forest*' – the student refers directly to evidence which is visible on the photo.

By meeting the descriptors partially, the answer is lower Level 2 and earns 3 marks. Notice that there are no statements which can be classed as judgements – which are essential for Level 3.

Marked sample answer 2

Judgement – the student makes a clear judgement at the start.

Evidence – observes one impact from the photo

Explanation – explains what the forest would be like before clearance (essential to understanding impacts)

Clearing rainforest areas like this can be disastrous. The forest cover has been cleared and burned. It used to shelter the soil so that when it rained heavily the rain would be intercepted and would drip slowly into the soil. Now if there's a storm, the rain will run off and will probably take the soil with it because there's nothing to protect it.

Soils in rainforests are infertile, so that clearing land for plantations may not be a good idea anyway. Probably the ash that's in the photo would be the only fertile part of the soil. So clearing forests is not likely to have any benefits for farmers.

Point – describes one impact (surface runoff)

Point – describes a second impact (soil erosion)

Explanation – explains why erosion occurs

Point – makes a point to support the next judgement, which is clarified

Judgement – another clear judgement about clearance

Evidence – the student observes the ash and uses this to support the point about soils

Judgement – a third judgement about clearance

Examiner feedback

The descriptor for Level 3 applies to this answer as follows:

- '*Provides a balanced discussion with well-developed impacts and detailed understanding of these impacts …*' – the student balances the answer with impacts about forest, rainfall runoff,, and soil erosion.
- '*…is able to make a judgement about these*'. The student makes three supported, well-evidenced judgements.
- '*Shows thorough identification of the evidence for the impacts of forest clearance, and understands the implications*' – the student uses evidence from the photo to apply to their answer.

By meeting these descriptors fully, the answer is top Level 3 and earns 6 marks.

On your marks

Using the command word 'Assess'

- **In this section you'll learn how to tackle 6-mark questions which use 'Assess' as a command word.**

Study **Figure 2**. It is a map that shows the distribution of Asian Indian British people in London, 2011.

Figure 2

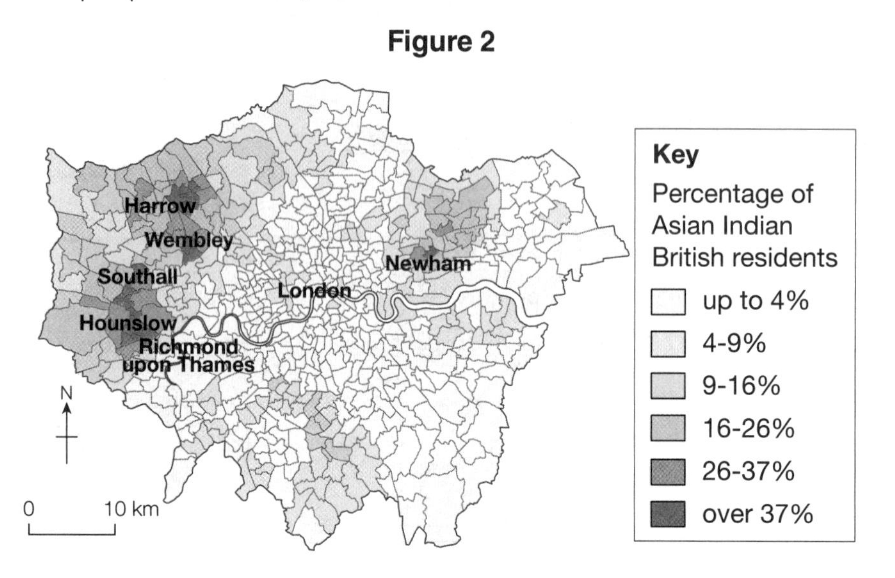

Assess the impacts of international migration on the growth and character of cities in the UK. Use **Figure 2** and your case study of a major city in the UK.

[6 marks]

1 Plan your answer

Before attempting to answer the question, remember to **BUG** it. Use the guidelines on page 21. Annotate it in the space below.

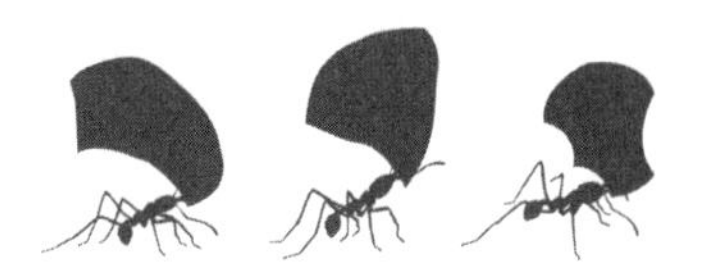

Assess the impacts of international migration on the growth and character of cities in the UK. Use **Figure 2** and your case study of a major city in the UK.

[6 marks]

Five steps to success!

The following five steps are used in this chapter to help you get the best marks.

1. **Plan your answer** – decide what to include and how to structure your answer.
2. **Write your answer** – use the answer spaces to complete your answer.
3. **Mark your answer** – use the mark scheme to self- or peer-mark your answer. You can also use this to assess sample answers in step 4 below.
4. **Sample answers** – sample answers are given to show you how to maximise marks for a question.
5. **Marked sample answers** – these are the same answers as for step 4, but are marked and annotated, so that you can compare these with your own answers.

PEEL your answer

Use PEEL notes to structure your answer. Refer to the guidelines on page 14 to help you.

Planning grid

Use this planning grid to help you write high-quality paragraphs. Remember to include links to show how your points relate to each other and to the question.

	PEE Paragraph 1	PEE Paragraph 2
Point		
Explanation		
Evidence *(from map or your own knowledge)*		
Link – a mini conclusion		

2 Write your answer

Assess the impacts of international migration on the growth and character of cities in the UK. Use **Figure 2** and your case study of a major city in the UK.

[6 marks]

Tip

Make a judgement!

Don't just describe and explain. If the question asks you to 'Assess', it wants you to make a judgement.

For example, 'Assess the impacts of rainforest clearance' would want you to say whether rainforest clearance is positive or negative, or how big the impacts have been (e.g. number of hectares cleared, or species affected).

One way of doing this is in a **mini-conclusion** – it need only be a sentence or two.

Strengths of the answer			
Ways to improve the answer			
Level		**Mark**	

3 Mark your answer

1. To help you to identify well-structured points in the answer, highlight or underline the:
   - points in red
   - explanations in orange
   - evidence in blue.
   - judgements by underlining. Judgements are important in order to reach Level 3 on questions with the command word is 'Assess'.
2. Use the mark scheme below to decide what mark to give. Six-mark questions are not marked using individual points, but instead you should choose a level and a mark based upon the quality of the answer as a whole.

Level	Marks	Description	Examples
3 (Detailed)	5–6	• Shows thorough understanding of the impacts of international migration on the growth and character of a named UK city. • Shows thorough application of knowledge and understanding in interpreting the map about the impacts of international migration.	• *The impacts of immigration have been great especially on the culture of UK cities. In Bradford, the Curry Mile attracts tourists as well as increasing the range of foods in the city.* • *Figure 2 shows that immigrants from particular countries, religions or cultures tend to live in areas close to each other, creating suburbs like Southall in west London.*
2 (Clear)	3–4	• Shows sound understanding of the impacts of international migration on the growth and character of a named UK city. • Shows sound application of knowledge and understanding in interpreting the map about the impacts of international migration.	• *Immigration has been the reason for half the recent growth of cities such as Bristol. In Bristol there are now over 50 languages spoken in the city.* • *Figure 2 shows that immigrants often settle in suburbs where there are cultural or ethnic groups like their own.*
1 (Basic)	1–2	• Shows limited understanding of the impacts of international migration on the growth and character of a named UK city. • Shows limited application of knowledge and understanding in interpreting the map about the impacts of international migration.	• *Bristol has a lot of immigrants living there so the city is growing.* • *Immigrants often live in the same sorts of areas, where they have their own shops or mosques, and they like living there.*
	0	No accurate response	

4 Sample answer

Read through the two sample answers 3 and 4 below.

a) Go through each one using the three colours on page 31 and remember to underline any **judgements**, because these are needed to meet the requirements of the command word 'Assess'.

b) Use the level descriptors to decide how many marks it is worth.

Question recap

Assess the impacts of international migration on the growth and character of cities in the UK. Use **Figure 2** and your case study of a major city in the UK.

Sample answer 3

Half of Bristol's population growth in recent years has been migrants from overseas, from countries such as Poland, because of the jobs available, such as in construction and the NHS. International migration has affected Bristol because people from over fifty countries have settled there. Over 6000 people live in Bristol who were born in Poland.

Like the map in Figure 2, many immigrants have changed the character of the parts of the city where they live, because they have their own shops or places of worship. This changes the culture in cities and there are festivals like the Notting Hill Carnival in London. So immigration has had a big effect on cities.

Strengths of the answer			
Ways to improve the answer			
Level		Mark	

Sample answer 4

Cities like Bristol are growing fast because of immigrants from other countries. There are jobs in Bristol which attract people to live there. It has meant that there is pressure on jobs and housing but the city gains because there are also new restaurants and festivals which help the city's image. When migrants arrive they look for work anywhere, but when they get jobs, their families come and join them, so that's what makes the population go up so quickly.

Strengths of the answer			
Ways to improve the answer			
Level		Mark	

5 Marked sample answer

Sample answers 3 and 4 are marked below. The text has been highlighted as follows to show how well each answer has structured points:

- points in red
- explanations in orange
- evidence in blue.
- judgements are underlined. These are important in order to reach Level 3 on questions whose command word is 'Assess'.

Marked sample answer 3

Half of Bristol's population growth in recent years has been migrants from overseas, from countries such as Poland, because of the jobs available, such as in construction and the NHS. International migration has affected Bristol because people from over fifty countries have settled there. Over 6000 people live in Bristol who were born in Poland.

Like the map in Figure 2, many immigrants have changed the character of the parts of the city where they live, because they have their own shops or places of worship. This changes the culture in cities and there are festivals like the Notting Hill Carnival in London. So immigration has had a big effect on cities.

Point – the student quantifies the amount of population growth due to immigration

Evidence – the example of Poland illustrates the point

Explanation – the reason is given for immigration, i.e. employment (with examples)

Point – student quantifies the extent of immigration from different countries

Evidence – the point is evidenced with the example of Poland

Explanation – the student explains how the character is changed, with examples

Point – the point helps to answer the part of the question dealing with changing character of cities

Evidence – student illustrates the point with an example

Judgement – the student makes a judgement about how migration changes the city. This is the weakest part of the answer

Examiner feedback

The descriptors for Level 3 applies to this answer as follows:

- '*Shows thorough understanding of the impacts of international migration on the growth and character of a named UK city*' – the student has been able to mention both the reasons for growth and the changing character of Bristol. A specific source country is named, and there is an example of the kind of cultural events resulting from immigration. It does not matter that these examples are chosen from different cities, as long as they are large and in the UK.
- '*Shows thorough application of knowledge and understanding in interpreting the map about the impacts of international migration*' – this is not so strong. Although the student refers to Figure 2, there is no real interpretation of what it shows.

By meeting the first descriptor fully, and the second one partly, the answer is low Level 3 in quality. The judgement is also weaker than would be needed for a top Level 3, so the answer is worth 5 marks.

Marked sample answer 4

Point – a general and non-specific point helps to answer the part of the question dealing with growth of cities

Explanation – the student briefly explains the reason for growth – but without specific examples

Cities like Bristol are growing fast because of immigrants from other countries. There are jobs in Bristol which attract people to live there. It has meant that there is pressure on jobs and housing but the city gains because there are also new restaurants and festivals which help the city's image. When migrants arrive they look for work anywhere, but when they get jobs, their families come and join them, so that's what makes the population go up so quickly.

Explanation – the student explains how jobs help to explain immigration but does not offer examples

Evidence – student evidences the point by showing the benefit of restaurants and festivals

Point – this point helps to answer the part of the question dealing with changing character of cities

Explanation – the student revisits the explanation for growth of population

Note that there are no judgements made in this answer.

Examiner feedback

The descriptor for Level 1 just about applies to this answer as follows:

- '*Shows sound understanding of the impacts of international migration on the growth and character of a named UK city*' – the student names Bristol and clearly understands how important migration, and immigration in particular, help to explain the city's rapid growth of population. In the latter part of the answer, the student also mentions the importance of family members as a reason for further increase. The student understands the impact of immigration in terms of food and festivals, but there are no specific examples.
- '*Shows limited application of knowledge and understanding in interpreting the map about the impacts of international migration*' – the student does not refer to Figure 2 at all in the answer.

By only partly meeting the descriptor for Level 2, the answer gains 3 marks.

On your marks

Using the command word 'Explain'

- **In this section you'll learn how to tackle 6-mark questions which use 'Explain' as a command word.**

Note that this question focuses upon coasts. If you have studied River landscapes and Glacial landscapes, but not Coastal landscapes, plan, write and mark the exam question as stated, but use a different photo.

- For River landscapes, use photo A on page 120 from the student book.
- For Glacial landscapes, use photo D on page 137 from the student book.

Study **Figure 3**. It is a photo that shows deposition of sediment along a stretch of coast in South Australia.

Figure 3

Using **Figure 3** and your own knowledge, explain how different landforms may be created by deposition of sediment.

[6 marks]

1 Plan your answer

Before attempting to answer the question, remember to **BUG** it. Use the guidelines on page 31. Annotate it in the space below.

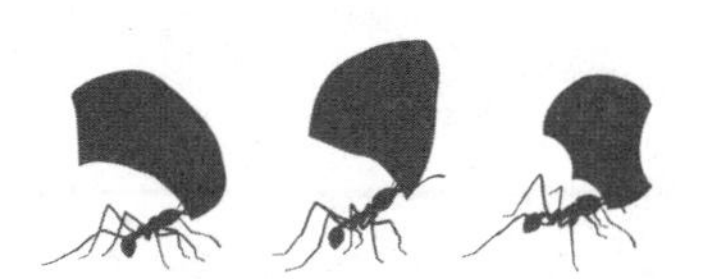

Using **Figure 3** and your own knowledge, explain how different landforms may be created by deposition of sediment.

[6 marks]

Five steps to success!

The following five steps are used in this chapter to help you get the best marks.

1. **Plan your answer** – decide what to include and how to structure your answer.
2. **Write your answer** – use the answer spaces to complete your answer.
3. **Mark your answer** – use the mark scheme to self- or peer-mark your answer. You can also use this to assess sample answers in step 4 below.
4. **Sample answers** – sample answers are given to show you how to maximise marks for a question.
5. **Marked sample answers** – these are the same answers as for step 4, but are marked and annotated, so that you can compare these with your own answers.

PEEL your answer

Use PEEL notes to structure your answer. Refer to the guidelines on page 14 to help you.

Planning grid

Use this planning grid to help you write high-quality paragraphs. Remember to include links to show how your points relate to each other and to the question.

	PEE Paragraph 1	PEE Paragraph 2
Point		
Explanation		
Evidence *(from photo or your own knowledge)*		
Link – a mini conclusion		

2 Write your answer

Using **Figure 3** and your own knowledge, explain how different landforms may be created by deposition of sediment.

[6 marks]

Tip

Make sure you explain!

Don't just describe. If the question asks you to 'Explain', it wants you to give reasons why something happens.

For example, 'Explain how a named landform is caused by erosion' would want you to say how erosion processes led to its formation, not simply a description of the landform.

Strengths of the answer			
Ways to improve the answer			
Level		**Mark**	

3 Mark your answer

1. To help you to identify well-structured points in the answer, highlight the:
   - points in red
   - explanations in orange
   - evidence in blue.
2. Use the mark scheme below to decide what mark to give.

Level	Marks	Description	Examples
3 (Detailed)	5–6	• Shows thorough application of knowledge and understanding to analyse information, giving detailed explanation of formation of coastal features. • Makes full analysis of the photo, using evidence to support the answer.	• *The coastal spit shown has been formed by longshore drift, caused by winds creating waves which hit the shore at an angle.* • *Figure 3 shows a coastal spit which has forced the river to divert from where it used to reach the sea.*
2 (Clear)	3–4	• Demonstrates specific and accurate knowledge of coastal processes and landforms. • Shows thorough understanding of the links between coastal processes and landforms.	• *Coastal spits are formed when waves break on the shore at an angle and take sediment along the coast, forming a long, sandy headland into the water.* • *Figure 3 shows how the river stops the spit from forming a bar which would join the two bits of coast together.*
1 (Basic)	1–2	• Demonstrates some knowledge of coastal processes and landforms. • Shows limited geographical understanding of the links between coastal processes and landforms.	• *The spit comes from waves which break on the beach, and longshore drift.* • *The photo shows a sandy beach which reaches almost across the river.*
	0	No accurate response	

4 Sample answer

Read through sample answer 5 below. Go through it using the three colours above and decide how many marks it is worth.

Question recap

Using **Figure 3** and your own knowledge, explain how different landforms may be created by deposition of sediment.

Sample answer 5

The photo shows a spit formed of sand washed up by waves on the beach. The waves approach at an angle and the swash takes sand up the beach, then it runs back down in a zigzag pattern. Then another wave picks it up, and deposits it further along the beach, and so on, forming an extension of land. The sand moves along the beach until it reaches a river, and the river current shapes it where the water runs out to sea. It looks like the river in the photo has had to divert around the spit.

Another landform formed by deposition is a sand bar, which is just like a spit except that there is no river to stop the movement of sand. The sand keeps on moving until it cuts off a lake or lagoon.

Strengths of the answer			
Ways to improve the answer			
Level		Mark	

5 Marked sample answer

Sample answer 5 is marked below. The text has been highlighted as follows to show how well the answer has structured points:

- points in red
- explanations in orange
- evidence in blue.

Marked sample answer 5

Point – the student names the landform and makes it clear it is formed of sand

Explanation – the process that begins to form the spit is described

The photo shows a spit formed of sand washed up by waves on the beach. The waves approach at an angle and the swash takes sand up the beach, then it runs back down in a zigzag pattern. Then another wave picks it up, and deposits it further along the beach, and so on, forming an extension of land, as shown in the photo. The sand moves along the beach until it reaches a river, and the photo shows how river current has shaped where the water runs out to sea. It looks like the river in the photo has had to divert around the spit.

Another landform formed by deposition is a sand bar, which is just like a spit except that there is no river to stop the movement of sand. The sand keeps on moving until it cuts off a lake or lagoon.

Evidence – the student evidences the process from the photo. This evidence is important when explaining processes as a sequence of stages

Evidence – further evidenced to include the part played by the river

Evidence – the point is further evidenced by explaining how the river is affected by the spit

Point – the student names a second depositional landform

Explanation – the process of bar formation is explained

Examiner feedback

This student gets a Level 3 and 5 marks… just! Most time is spent explaining the spit, and less on a second landform. It would be better to split time equally between two, or three landforms.

That said, the descriptors for Level 3 applies to this answer as follows:

- '*Shows thorough application of knowledge and understanding* …' – the student describes spit formation in detail.
- '*… giving detailed explanation of formation of coastal features*' – the student describes spit formation as a sequence of events. The difference between a spit and a bar is explained.
- '*Makes full analysis of the photo, using evidence to support the answer*' – the student recognises and names the landform, and explains the impact of the spit on the river.

On your marks

Using the command word 'Evaluate'

- **In this section you'll learn how to tackle 9-mark questions that use 'Evaluate' as a command word.**

Nine-mark questions – what's different?

Nine-mark questions are marked using three levels, but they differ from 6-mark questions.

- All three levels in the mark scheme require you to write more and to a higher standard. This is reflected in the marks – Level 1 ranges from 1 to 3 marks, Level 2 from 4 to 6 marks, and Level 3 from 7 to 9 marks.
- Command words can be more demanding – they use 'Justify' and 'Evaluate', as well as commands such as 'Discuss' and 'Assess'.
- As with 6-mark questions, you could be given a statement, but 9-mark questions may ask you to decide whether you agree or disagree with it (as in the example below), by developing a supporting argument.

A **6-mark question** might ask:

'Volcanic eruptions can cause important changes to the global climate'.
Use evidence to support this statement.

A **9-mark question** might ask:

'Natural hazards generally affect vulnerable people most seriously'.
To what extent do you agree with this statement?

How is SPaG assessed?

One 9-mark question on each of Papers 1 and 2 will assess your accuracy of spelling, punctuation, grammar, and the use of specialist terminology (known as SPaG). In each of these questions, 3 marks are allocated as follows:

- High performance – 3 marks
- Intermediate performance – 2 marks
- Threshold performance – 1 mark.

Examiners mark SPaG based on your:

- spelling accuracy
- punctuation – the use of commas, full stops and semi-colons. Try reading an answer aloud and see if it leaves you gasping for breath! If it does, it needs more punctuation!
- paragraphing.

Evaluate the evidence that suggests that global climate is currently changing.

[9 marks] [+ 3 SPaG marks]

Five steps to success!

The following five steps are used in this chapter to help you get the best marks.

1. **Plan your answer** – decide what to include and how to structure your answer.
2. **Write your answer** – use the answer spaces to complete your answer.
3. **Mark your answer** – use the mark scheme to self- or peer-mark your answer. You can also use this to assess sample answers in step 4 below.
4. **Sample answers** – sample answers are given to show you how to maximise marks for a question.
5. **Marked sample answers** – these are the same answers as for step 4, but are marked and annotated, so that you can compare these with your own answers.

1 Plan your answer

Before attempting to answer the question on page 40, remember to **BUG** it.

✓ **Box** the command word.
✓ **Underline** the following:
- the theme
- the focus
- any evidence required
- the number of examples needed.

✓ **Glance** back over the question – to make sure you include everything in your answer.

Use the BUG below to plan your own answer.

> **Tip**
> **'Evaluate' means stating how strong each piece of evidence is**
> Don't just describe and explain. 'Evaluate' needs you to show if the research and evidence is strong or not. For example, evidence of global climate change might be shown by shrinking glaciers. Then draw it together in a **mini-conclusion**.

Command word: Evaluate means 'judge on its strengths and weaknesses'. You need to decide whether evidence is strong or weak.

Evidence: Support your answer with evidence from your own knowledge and understanding, such as shrinking glaciers and seasonal weather changes. developed.

Evaluate the evidence that suggests that global climate is currently changing.
[9 marks] [+ 3 marks SPaG]

Focus and number of examples: The focus is evidence for a changing global climate. For a 9-mark question, you need three points which are well developed. Each piece of evidence needs to be written in a paragraph. You need a mini conclusion.

Theme: Climate change is linked to the theme The challenge of natural hazards, assessed in Paper 1, Section A of your exam. The question is compulsory.

SPaG: The answer needs to be planned and organised into paragraphs, and written in sentences. Check your spelling and punctuation.

PEEL your answer

Use PEEL notes to structure your answer. This will help you to communicate your ideas to the examiner in the clearest way. PEEL has four stages:

- **P**oint – give at least three pieces of evidence for this question. Use sentences, not bullet points.
- **E**xplain – give reasons for each piece of evidence and how it shows climate is changing. Use sentence starters such as: 'This is because ...', 'One reason is ...'.
- **E**vidence – include details from named examples to support your evidence.
- **L**ink – link the pieces of evidence to each other. Use PEE sentence starters such as: 'A second piece of evidence is ....' or 'Secondly ...'. Finish it with a one-sentence conclusion about how likely it is that climate is changing.

Planning grid

Use this planning grid to help you write high-quality paragraphs. Remember to include links to show how your points relate to each other and to the question. Note that this is a 9-mark question, so needs three PEE Paragraphs.

Note that this is different from the 6-mark questions, because the fourth row helps you focus on the word 'evaluate'. Remember in this question to ***evaluate the evidence***.

	PEE Paragraph 1	PEE Paragraph 2	PEE Paragraph 3
Point			
Explanation			
Evidence *(from photo or your own knowledge)*			
Evaluation of the evidence and mini-conclusion			

2 Write your answer

Evaluate the evidence that suggests that global climate is currently changing.

[9 marks] [+ 3 SPaG marks]

Strengths of the answer			
Ways to improve the answer			
Level		**Mark out of 9**	
SPaG level		**Mark out of 3**	

3 Mark your answer

1. To help you to identify well-structured points in the answer, highlight the:
   - points in red • explanations in orange • evidence in blue.

2. Use the mark scheme below to decide what mark to give. Nine-mark questions are not marked using individual points, but instead you should choose a level and a mark based upon the quality of the answer as a whole.

Level	Marks	Description	Examples
3 (Detailed)	7–9	• Shows detailed knowledge of the evidence for changing climate.	• *IPCC research shows that average global sea level has risen by 10–20 cm since 1920.*
		• Shows thorough geographical understanding of the processes by which climate may be changing globally.	• *This is probably due to rising global temperatures which melt ice caps, from which more water goes into the sea.*
		• Shows application of knowledge and understanding in a coherent and reasoned way in evaluating the evidence for climate change.	• *This is likely to be reliable evidence as the IPCC consists of thousands of the world's best scientists.*
2 (Clear)	4–6	• Shows clear knowledge of the evidence for changing climate.	• *Scientists show that global sea levels have risen in the past 100 years.*
		• Shows some geographical understanding of the processes by which climate may be changing globally.	• *This is due to global warming which increases temperatures and melted ice caps and glaciers which go into the sea.*
		• Shows reasonable application of knowledge and understanding in evaluating the evidence for climate change.	• *We know sea level is rising because countries with coastlines are getting flooded.*
1 (Basic)	1–3	• Shows limited knowledge of the evidence for changing climate.	• *World temperatures are increasing all the time and winters are getting warmer.*
		• Shows slight geographical understanding of the processes by which climate may be changing globally.	• *Global warming is making the seasons different and there are more floods.*
		• Shows limited application of knowledge and understanding in evaluating the evidence for climate change.	• *Scientists think more floods and storms are because of global warming.*
	0	No accurate response	

Table for sample answer 2 (opposite)

Strengths of the answer			**Ways to improve the answer**		
Level		**Mark**	**SPaG Level**		**Mark**	

4 Sample answers

Read through these two sample answers.

a) Go through each one using the three colours on page 44.
b) Use the level descriptors to decide how many marks it is worth.
c) Remember to give a SPaG mark.

Question recap

Evaluate the evidence that suggests that global climate is currently changing.

Sample answer 1

Many sources of evidence show how climate is changing. Temperatures have risen globally since the nineteenth century by about 0.8 °C. This is probably due to carbon emissions of greenhouse gases like CO_2 from burning of fossil fuels.

Temperatures seem to be getting warmer all the time, so that sea level will carry on rising. Already some islands in the Pacific have been flooded and countries like Bangladesh have severe floods because much of the country is very low-lying. Glaciers in mountains like the Himalayas have been melting because temperatures are rising, so that this all goes to the sea via rivers and makes sea level rise.

Another piece of evidence is that the seasons seem to be changing, so that spring is earlier, and winters are not as cold as they were, and have less snow. Birds now migrate earlier than they did and their nests are being built nine days earlier than forty years ago. So that all seems to mean that there is a lot of evidence that climate is changing.

Strengths of the answer				Ways to improve the answer			
Level		Mark		SPaG Level		Mark	

Sample answer 2

Globally the climate is warming, with evidence to prove that this is the case. Global temperatures are 1 °C warmer than they were 100 years ago because greenhouse gas emissions have increased. It is hard to know exactly what temperatures were like in 1900 and more people and organisations record the weather now than at that time, but there were thermometers, just fewer of them. So, some of the evidence could be questionable, just because there are more recordings.

Even if temperature recordings are not completely reliable, there is a lot of evidence to show that sea level is rising globally by about 20 cm in 100 years, partly because ocean water expands when it warms and so it rises. Many coastal areas are flooding more now, so it is a global process and not just evidence from one place.

Other evidence which shows that temperatures are rising comes from retreating glaciers and ice sheets because they are melting. Many glaciers have been photographed for over 100 years, and many in the Alps and on Greenland show that they have retreated a long way from where they were.

See opposite for table for sample answer 2

5 Marked sample answers

Sample answers 1 and 2 are marked below. The text has been highlighted as follows to show how well each answer has structured points:

- points in red
- explanations in orange
- evidence in blue.
- evaluations are underlined. These are important to reach Level 3 on questions whose command word is 'Evaluate'.

Marked sample answer 1

Explanation – a reason is given for warming of the global climate

Point – the student quantifies the amount of warming

Many sources of evidence show how climate is changing. Temperatures have risen globally since the nineteenth century by about 0.8 °C. This is probably due to carbon emissions of greenhouse gases like CO_2 from burning of fossil fuels.

Temperatures seem to be getting warmer all the time, so that sea level will carry on rising. Already some islands in the Pacific have been flooded and countries like Bangladesh have severe floods because much of the country is very low-lying. Glaciers in mountains like the Himalayas have been melting because temperatures are rising, so that this all goes to the sea via rivers and makes sea level rise.

Another piece of evidence is that the seasons seem to be changing, so that spring is earlier, and winters are not so cold as they were and have less snow. Birds now migrate earlier than they did and their nests are being built nine days earlier than forty years ago. So that all seems to mean that there is a lot of evidence that climate is changing.

Point – the student offers further evidence

Explanation – a reason is given for flooding in many countries

Point – further evidence is given for climate change

Explanation – a reason is given for glaciers melting

Evidence – the student discusses retreating glaciers as evidence for rising sea levels

Point – further evidence is given for climate change (winters less cold)

Evidence – the point is extended using bird migration as evidence

Examiner feedback

Examiners see many answers of this kind. This student knows a lot and has learnt facts and figures. The answer is a problem, though, because there is no evaluation. The student needs to ask themselves – '*what's the evidence that glaciers are melting, and is it reliable? How do I know it's reliable?*'.

The answer is therefore a mix of levels:

- Almost Level 3 for knowledge about climate change and warming.
- Explanations are mid-Level 2 because they do not always link to the warming climate (e.g. flooding in Bangladesh is explained because it is low lying, not because of sea level change).
- However, there is no evaluation.

Faced with this, examiners have to do a 'best fit' or a kind of average. The examiner gives this low Level 2 overall of 4 marks.

Marked sample answer 2

Explanation – the student briefly explains the increase

Point – the student makes the point about increasing temperatures

Globally the climate is warming, with evidence to prove that this is the case. Global temperatures are 1 °C warmer than they were 100 years ago because greenhouse gas emissions have increased. It is hard to know exactly what temperatures were like in 1900 and more people and organisations record the weather now than at that time, but there were thermometers, just fewer of them. So, some of the evidence could be questionable, just because there are more recordings.

Even if temperature recordings are not completely reliable, there is a lot of evidence to show that sea level is rising globally by about 20 cm in 100 years, partly because ocean water expands when it warms and so it rises. Many coastal areas are flooding more now, so it is a global process and not just evidence from one place.

Other evidence which shows that temperatures are rising comes from retreating glaciers and ice sheets because they are melting. Many glaciers have been photographed for over 100 years, and many in the Alps and on Greenland show that they have retreated a long way from where they were.

Evaluation – the student gives one reason why temperature readings may not be accurate

Evaluation – the student extends the evaluation by referring to volume of temperature recordings

Point – the student makes a second point about rising sea level

Explanation – the student gives a reason for this

Evaluation – the student shows that this is probably reliable as many places experience the same thing

Point – the student makes the point about retreating glaciers

Evaluation – the student refers to the reliability of photos taken over a long time to show change

Explanation – the student explains this

Examiner feedback

This is a top quality answer which was given the full 9 marks.

Notice that this student has shown less knowledge and understanding than the student in sample answer 1 opposite, but that nearly half of the answer is spent in showing whether the evidence for change is reliable or not. That's what you need to do in a question whose command word is 'Evaluate'. Spend as much time on evaluating as the time you spend in showing your knowledge and understanding.

For SPaG, the student was awarded the full 3 marks. The answer is in paragraphs, is well spelt, and punctuation gives meaning to what the student is saying.

On your marks

Using the command word 'Justify'

- **In this section you'll learn how to tackle 9-mark questions which use 'Justify' as a command word.**

Study **Figure 1**. It shows Rocinha favela, a low-income squatter settlement in Rio de Janeiro.

Figure 1

'For those who live in low-income areas of cities such as Rio de Janeiro, life presents far more problems than benefits'.

Do you agree with this statement? YES ☐ NO ☐

Justify your decision, using **Figure 1** and your case study of a major city in an LIC or NEE.

[9 marks] [+ 3 SPaG marks]

1 Plan your answer

Before attempting to answer the question, remember to **BUG** it. Use the guidelines on page 21 and annotate it below.

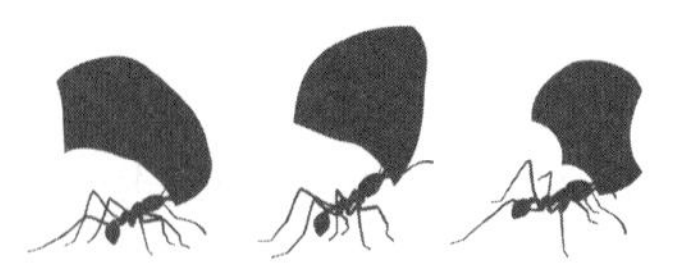

'For those who live in low income areas of cities such as Rio de Janeiro, life presents far more problems than benefits'.

Do you agree with this statement? YES ☐ NO ☐

Justify your decision, using **Figure 1** and your case study of a major city in an LIC or NEE.

[9 marks] [+ 3 SPaG marks]

Five steps to success!

The following five steps are used in this chapter to help you get the best marks.

1. **Plan your answer** – decide what to include and how to structure your answer.
2. **Write your answer** – use the answer spaces to complete your answer.
3. **Mark your answer** – use the mark scheme to self- or peer-mark your answer. You can also use this to assess sample answers in step 4 below.
4. **Sample answers** – sample answers are given to show you how to maximise marks for a question.
5. **Marked sample answers** – these are the same answers as for step 4, but are marked and annotated, so that you can compare these with your own answers.

PEEL your answer

Use PEEL notes to structure your answer.
Use the guidelines on page 14 to help you.

Justify means supporting your decision with evidence!

Don't just describe and explain. Use data or examples (including the photo) to show that your decision is correct, and say briefly why you rejected the alternative. You should state clearly what your answer is, and draw it together at the end in a **mini-conclusion** of one or two sentences.

Planning grid

Use this planning grid to help you write high-quality paragraphs. Remember to include links to show how your points relate to each other and to the question.

Note that the fourth row helps you focus on the command 'justify'. Remember, you must ***justify your decision.***

	PEE Paragraph 1	PEE Paragraph 2	PEE Paragraph 3
Point			
Explanation			
Evidence *(from photo or your own knowledge)*			
How this supports your decision and mini-conclusion			

2 Write your answer

'For those who live in low-income areas of cities such as Rio de Janeiro, life presents far more problems than benefits'.

Do you agree with this statement? YES ☐ NO ☐

Justify your decision, using **Figure 1** and your case study of a major city in an LIC or NEE.

[9 marks] [+ 3 SPaG marks]

Strengths of the answer			
Ways to improve the answer			
Level		**Mark out of 9**	
SPaG level		**Mark out of 3**	

3 Mark your answer

1. To help you to identify well-structured points in the answer, highlight or underline the:
   - points in red • explanations in orange • evidence in blue.
   - points which show the student is justifying (or supporting) the statement by underlining. These might support one side of the argument, or balance it before reaching a conclusion.
2. Use the mark scheme below to decide what mark to give. Nine-mark questions are not marked using individual points, but choose a level and a mark based upon the quality of the answer as a whole.

Level	Marks	Description	Examples
3 (Detailed)	7–9	• Shows comprehensive and specific knowledge of the problems and benefits of one or more cities.	• *In Rio, a third of homes have no electricity (or have illegal hook-ups from wires, as in the photo) and half have no sewerage connections.*
		• Shows thorough and accurate understanding of the problems and benefits of one or more cities.	• *One reason is that favelas like Rocinha are growing so quickly that the city council cannot keep pace with population growth.*
		• Shows effective application of knowledge and understanding in making a judgement and reaching a substantiated conclusion. Justification is detailed and balanced.	• *This shows the statement is true because electricity and sewerage connection are basics for a reasonable life. But there are benefits, such as provision of schooling.*
2 (Clear)	4–6	• Shows reasonable knowledge of the problems and benefits of one or more cities.	• *Cities like Rio often have no water or sewerage connections, and electricity in the photo looks unsafe too.*
		• Shows clear geographical understanding of the problems and benefits of one or more cities.	• *This is because people are poor and cannot afford water or electricity bills.*
		• Shows reasonable application of knowledge and understanding in making a judgement and reaching a conclusion. Justification is clear and well supported.	• *So the statement is true because most people do not have a decent lifestyle with basics that we would take for granted.*
1 (Basic)	1–3	• Shows limited knowledge of the problems and benefits of one or more cities. Answers may be largely generic.	• *Developing cities have no water or sewerage connection and many health problems from drinking bad water.*
		• Shows some geographical understanding of the problems and benefits of living in cities.	• *There are so many people that the city cannot keep pace with them all.*
		• Shows limited application of knowledge and understanding in making a judgement and/or reaching a conclusion. Justification is limited to one or more simple points.	• *So the statement is correct because life there is very hard and the city cannot support all those people.*
	0	No accurate response	

Remember to give a mark for SPaG too! The criteria can be found on page 40.

4 Sample answers

Read through the two sample answers 3 and 4 below.

a) Go through it using the three colours in on page 51, and underline parts that meet the requirements of the command word 'Justify'.

b) Use the level descriptors to decide how many marks it is worth.

c) Remember to give a SPaG mark.

Do you agree with the statement on page 50. Justify your decision, using **Figure 1** and your case study of a major city in an LIC or NEE.

Sample answer 3

I agree with the statement. Rio's favelas are growing so quickly that it is hard to keep pace with services needed like water. Rocinha has grown three times its size since 2010. It is better than it was because now houses are being built out of brick instead of timber and odd bits of metal, and they also have water and electricity. There are shops there and many services like health facilities that you would expect. But I agree with the statement because Rocinha is probably one of Rio's best favelas and there are many worse that do not have half the benefits that it has. You wouldn't choose to live there if you had more money so areas like that are still for low income people, so I still think the statement is true.

Elsewhere Rio has squatter settlements, which are places where people just put together their own shacks illegally. Some shacks are on sloping land because nobody else wants to live there and are a long way from jobs in the city centre. But when it rains heavily, people are vulnerable, because in 2010 over 200 people were killed in a landslide. This shows the statement is true, because the poor have to live there – people with jobs and decent incomes would never choose to.

Strengths of the answer			**Ways to improve the answer**		
Level		**Mark**	**Level**		**Mark**

Sample answer 4

I don't agree with the statement. It is true that people living in squatter settlements have a lot of problems like they don't have water supply or sewerage connections and when you walk down the street in the photo then you might be electrocuted as the wires don't look very safe. But cities have many jobs for people and so the people who have moved there from the countryside are often employed more than if they had stayed in rural areas. Many rural areas do not have schools and cities like Rio have plenty of schools for all ages maybe universities too and often hospitals and medical treatment in cities that you don't have in the countryside. So it's not perfect living in Rio but it can be better than a lot of places so I don't agree with the statement.

Strengths of the answer			**Ways to improve the answer**		
Level		**Mark**	**Level**		**Mark**

5 Marked sample answers

Sample answers 3 and 4 are marked below. The text has been highlighted as follows to show how well each answer has structured points:

- points in red • explanations in orange • evidence in blue.
- justifications are underlined. These are important in order to reach Level 3 on questions whose command word is 'Assess'.

Marked sample answer 3

I agree with the statement. Rio's favelas are growing so quickly that it is hard to keep pace with services needed like water. Rocinha has grown three times its size since 2010. It is better than it was because now houses are being built out of brick instead of timber and odd bits of metal, and they also have water and electricity. There are shops there and many services like health facilities that you would expect. But I agree with the statement because Rocinha is probably one of Rio's best favelas and there are many worse that do not have half the benefits that it has. You wouldn't choose to live there if you had more money so areas like that are still for low income people, so I still think the statement is true.

Elsewhere Rio has squatter settlements, which are places where people just put together their own shacks illegally. Some shacks are on sloping land because nobody else wants to live there and are a long way from jobs in the city centre. But when it rains heavily, people are vulnerable because in 2010 over 200 people were killed in a landslide. This shows the statement is true, because the poor have to live there – people with jobs and decent incomes would never choose to.

Point – the student quantifies the growth of Rocinha

Explanation – an explanation of the impact of Rocinha's growth

Evidence – the student uses the evidence of building materials

Evidence – the student uses further evidence of shops and health services

Justification – the student makes a comparison to justify their choice

Justification – the student makes a further statement to justify their choice

Point – the student mentions squatter settlements

Explanation – squatter settlements are explained

Evidence – the student uses evidence about land used by squatter settlements

Point – the student describes the vulnerability of people

Explanation – an explanation is given to illustrate this

Justification – the student gives one further supporting statement to justify their choice, though it is very similar to the second justification

Examiner feedback

This student knows a lot and has a clear view of what living in a favela might be like. The points are well made and the extended points offer a lot of detail about favelas to support the answer. The student also justifies clearly why they have reached an opinion. The answer is generally Level 3 in quality. However, it is not perfect, as the student doesn't directly refer to the photo and is short on any data. It is Level 3 and the justification is sound, so it earns 8 marks.

For SPaG, the student was awarded 3 marks. The answer is in paragraphs, is well spelt, and punctuated accurately.

Marked sample answer 4

Point – the student makes an illustrated point about squatter settlements

Evidence – this point is extended by using evidence from the photo about electricity

I don't agree with the statement. It is true that people living in squatter settlements have a lot of problems like they don't have water supply or sewerage connections and when you walk down the street in the photo then you might be electrocuted as the wires don't look very safe. But cities have many jobs for people and so the people who have moved there from the countryside are often employed more than if they had stayed in rural areas. Many rural areas do not have schools and cities like Rio have plenty of schools for all ages maybe universities too and often hospitals and medical treatment in cities that you don't have in the countryside. So it's not perfect living in Rio but it can be better than a lot of places so I don't agree with the statement.

Point – the student makes the point about employment in cities

Evidence – the student uses evidence of employment to compare cities and rural areas

Point – the student makes the point about education in cities

Justification – the student makes a single statement about living in Rio, though this is not a quality comparison.

Evidence – the student uses the evidence of health care to extend the point

Examiner feedback

This is a medium quality answer which was given 4 marks in the middle of Level 2. The student makes three valid points about living in squatter settlements and extends it with some detail, but a named city occurs just once in the last sentence. Generic writing – without naming a place – is normally typical of Level 1, so the student has saved themselves by naming Rio twice. The level of justification is weak; there is no other named place to compare Rio with, simply mentioning rural areas.

This student probably knows more than this, so some revision of a named city would have earned the student higher marks, perhaps with some named examples of parts of Rio, or some data illustrating households with water supply etc. Justification needs to be more than just a general statement at the end.

For SPaG, the student was awarded 1 mark. The answer is in a single paragraph, and although it is well spelt, there is a lack of punctuation.
This makes it harder to read and make sense of.

On your marks

Using the command 'To what extent'

- **In this section you'll learn how to tackle 9-mark questions that use the command 'To what extent'.**

Study **Figure 2**, a table showing development indicators for Nigeria, 1995 and 2015.

Figure 2

	1995	2015
GNI total (US$) in PPP	200 billion	1 trillion
GNI per capita (US$) in PPP	1850	5900
% people working in agriculture	54	70
% people working in industry	10	10
Exports value (US$)	11.9 billion	45.9 billion
Imports value (US$)	8.3 billion	52.3 billion
Unemployment rate %	28	13.9
Living in poverty %	36	70
Main exports	Petroleum and petroleum products (95% of total), cocoa, rubber	Petroleum and petroleum products (95% of total), cocoa, rubber
Main Imports	Machinery and equipment, manufactured goods, food and animals	Machinery, chemicals, transport equipment, manufactured goods, food and live animals

(Source: CIA Factbook)

To what extent do the data in **Figure 2**, and your own knowledge, suggest that Nigeria is a typical newly emerging economy (NEE)?

[9 marks] [+ 3 SPaG marks]

Five steps to success!

The following five steps are used in this chapter to help you get the best marks.

1. **Plan your answer** – decide what to include and how to structure your answer.
2. **Write your answer** – use the answer spaces to complete your answer.
3. **Mark your answer** – use the mark scheme to self- or peer-mark your answer. You can also use this to assess sample answers in step 4 below.
4. **Sample answers** – sample answers are given to show you how to maximise marks for a question.
5. **Marked sample answers** – these are the same answers as for step 4, but are marked and annotated, so that you can compare these with your own answers.

1 Plan your answer

Before attempting to answer the question, remember to **BUG** it. Use the guidelines on page 21 and annotate it below.

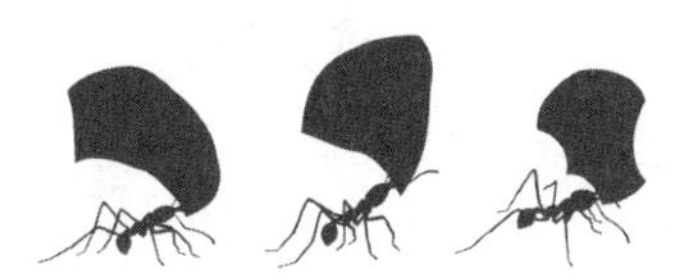

To what extent do the data in **Figure 2**, and your own knowledge, suggest that Nigeria is a typical newly emerging economy (NEE)?

[9 marks] [+ 3 SPaG marks]

PEEL your answer

Use PEEL notes to structure your answer.
Use the guidelines on page 14 to help you.

Planning grid

Use this planning grid to help you write high-quality paragraphs. Remember to include links to show how your points relate to each other and to the question.

Note that the fourth row helps you focus on the phrase '**to what extent**'. Remember, you must ***find evidence on both sides***!

Tip

How to answer 'To what extent' questions

Don't just describe. If the question asks you to explain 'to what extent', it means that you should be able to see ways in which something does happen, and ways it does not. For example, the data in Figure 2 have some data which suggest that Nigeria is an NEE and some that do not.

	PEE Paragraph 1	PEE Paragraph 2	PEE Paragraph 3
Point			
Explanation			
Evidence *(from Figure 2 or your own knowledge)*			
Does this suggest that Nigeria is an NEE or not? Include a mini-conclusion			

2 Write your answer

To what extent do the data in **Figure 2**, and your own knowledge, suggest that Nigeria is a typical newly emerging economy (NEE)?

[9 marks] [+ 3 SPaG marks]

Strengths of the answer			
Ways to improve the answer			
Level		**Mark out of 9**	
SPaG level		**Mark out of 3**	

3 Mark your answer

1. To help you to identify well-structured points in the answer, highlight or underline the:
   - points in red • explanations in orange • evidence in blue.
   - parts where '<u>to what extent</u>' is addressed are underlined.

2. Use the mark scheme below to decide what mark to give. Nine-mark questions are not marked using individual points, but instead choose a level and a mark based upon the quality of the answer as a whole.

Level	Marks	Description	Examples
3 (Detailed)	7–9	• Shows a comprehensive and accurate knowledge of the country, and the characteristics of an NEE.	• *Nigeria's GNI has grown rapidly by five times between 1995 and 2015.*
		• Shows a thorough geographical understanding of the country and the characteristics of an NEE.	• *Its oil has enabled it to grow rapidly because demand for oil has helped other NEEs and major economies to keep growing.*
		• Shows a thorough application of knowledge and understanding in evaluating the extent to which the country is or is not an NEE.	• *Rapid economic growth is typical of NEEs, so Nigeria conforms to that, but it still relies almost completely on oil for its exports, which is more like an LIC which depends on just one industry.*
2 (Clear)	4–6	• Shows reasonable knowledge of the country, and the characteristics of an NEE.	• *Nigeria's GNI has grown rapidly between 1995 and 2015.*
		• Shows clear geographical understanding of the country and the characteristics of an NEE.	• *Its oil is the reason for its growth because it has some of the world's largest reserves.*
		• Shows reasonable application of knowledge and understanding in evaluating the extent to which the country is or is not an NEE.	• *Its rapid growth is typical of NEEs.*
1 (Basic)	1–3	• Shows limited knowledge of the country, and the characteristics of an NEE.	• *Nigeria's economy has grown since 1995.*
		• Shows some geographical understanding of the country and the characteristics of an NEE.	• *It still sells a lot of oil which helps it to afford its imports. Developing countries have to have something to export to develop.*
		• Shows limited application of knowledge and understanding in evaluating the extent to which the country is or is not an NEE.	• *Nigeria's oil has helped it to develop since 1995 so its economy has grown.*
	0	No accurate response	

Remember to give a mark for SPaG too! The criteria can be found on page 40.

4 Sample answers

Read through sample answer 5 below.

a) Go through it using the three colours in on page 58, and underline parts that meet the requirements of the command 'To what extent'.
b) Use the level descriptors to decide how many marks it is worth.
c) Remember to give a SPaG mark.

Question recap

To what extent do the data in **Figure 2**, and your own knowledge, suggest that Nigeria is a typical newly emerging economy (NEE)?

Sample answer 5

In the table, Nigeria's economy has grown by five times in GNI between 1995 and 2015, and its GNI per capita has grown by about three and a half times. Its exports have also gone up, though its imports have gone up by more, so perhaps it is more in debt than it was in 1995 because now it does not earn enough to pay for its imports. The imports are machinery and transport equipment, which have probably gone up a lot in price. Unemployment has halved, which is good for the people of Nigeria though the number living in poverty has really gone up, and I would not expect that in an NEE. In NEEs industry is supposed to give people jobs and so unemployment goes down.

So I would say that Nigeria is an NEE because it is now earning a lot more than it was, but it looks like poverty in increasing.

Strengths of the answer			
Ways to improve the answer			
Level		**Mark out of 9**	
SPaG level		**Mark out of 3**	

Marked sample answers

Sample answer 5 is marked below. The text has been highlighted as follows to show how well the answer has structured points:

- points in red • explanations in orange • evidence in blue.
- parts where 'to what extent?' are addressed are underlined.

Marked sample answer 5

Point – the student identifies that Nigeria's economy has grown

Evidence – the student identifies data increase in GNI

Evidence – the student further evidences increase in GNI per capita

In the table, Nigeria's economy has grown by five times in GNI between 1995 and 2015, and its GNI per capita has grown by about three and a half times. Its exports have also gone up, though its imports have gone up by more, so perhaps it is more in debt than it was in 1995 because now it does not earn enough to pay for its imports. The imports are machinery and transport equipment, which have probably gone up a lot in price. Unemployment has halved, which is good for the people of Nigeria though the number living in poverty has really gone up, and I would not expect that in an NEE. In NEEs industry is supposed to give people jobs and so unemployment goes down.

So I would say that Nigeria is an NEE because it is now earning a lot more than it was, but it looks like poverty is increasing.

Point – the student identifies that Nigeria's exports have increased

Evidence – the student identifies greater increase in imports

Explanation – the student explains the debt that could arise if imports increase faster than exports

Evidence – the student identifies examples of imports

Evidence – the student has identified that poverty has increased by more

Evidence – the student identifies that unemployment has halved

To what extent – the student shows understanding of how this would not be expected in an NEE

To what extent – the conclusion rounds off two points to answer the question – but it might have been better if it was done each time the student identifies changes in the data

Examiner feedback

This student gets a low Level 3 and was given 7 marks. The student is clearly very good with data – six evidenced points have been correctly made about changes in Nigeria between 1995 and 2015, but for a top Level 3 there needs to be more explanation. The student has also left any reference to the extent to which Nigeria is an NEE until the last part of the answer. To reach the top of Level 3, the student should have made this evaluation each time a change in data had been identified. The student doesn't make it clear that rapid economic growth is expected in an NEE, but does state that you would not expect increased percentages of people living in poverty.

For SPaG, the student was awarded 3 marks. The answer is in paragraphs, is well spelt, and punctuated accurately.

GCSE 9-1 Geography AQA
Practice Paper

Paper 1 Living with the physical environment

Time allowed: 1 hour 30 minutes
Total number of marks: 88 (including 3 marks for spelling, punctuation, grammar and specialist terminology (SPaG))

Instructions
Answer **all** questions in Section A and Section B
Answer **two** questions in Section C

Section A The challenge of natural hazards

Answer **all** questions in this section.

Question 1 The challenge of natural hazards

0 1 · 1 What is a natural hazard?

[1 mark]

A natural hazard is an event that occurs without human interference but will cause harm to either humans or their property

0 1 · 2 Give **one** example of each of the following types of natural hazard.

Atmospheric	Geological/tectonic	Hydrological
Tropical storm	Earthquake	~~[illegible]~~ Flooding

[3 marks]

0 1 · 3 Explain why people continue to live in areas that are at risk from a geological or tectonic hazard.

[4 marks]

As often due to these hazards the land may be cheaper and hazards ~~often~~ don't occur very ~~particularly~~ frequently. Volcanic activity can create fertile land tourism causes these areas to benefit greatly. As a result, many find that the benefit of cheap housing outweighs the risks of natural hazards.

Study **Figure 1**, a diagram of the global atmospheric circulation

Figure 1

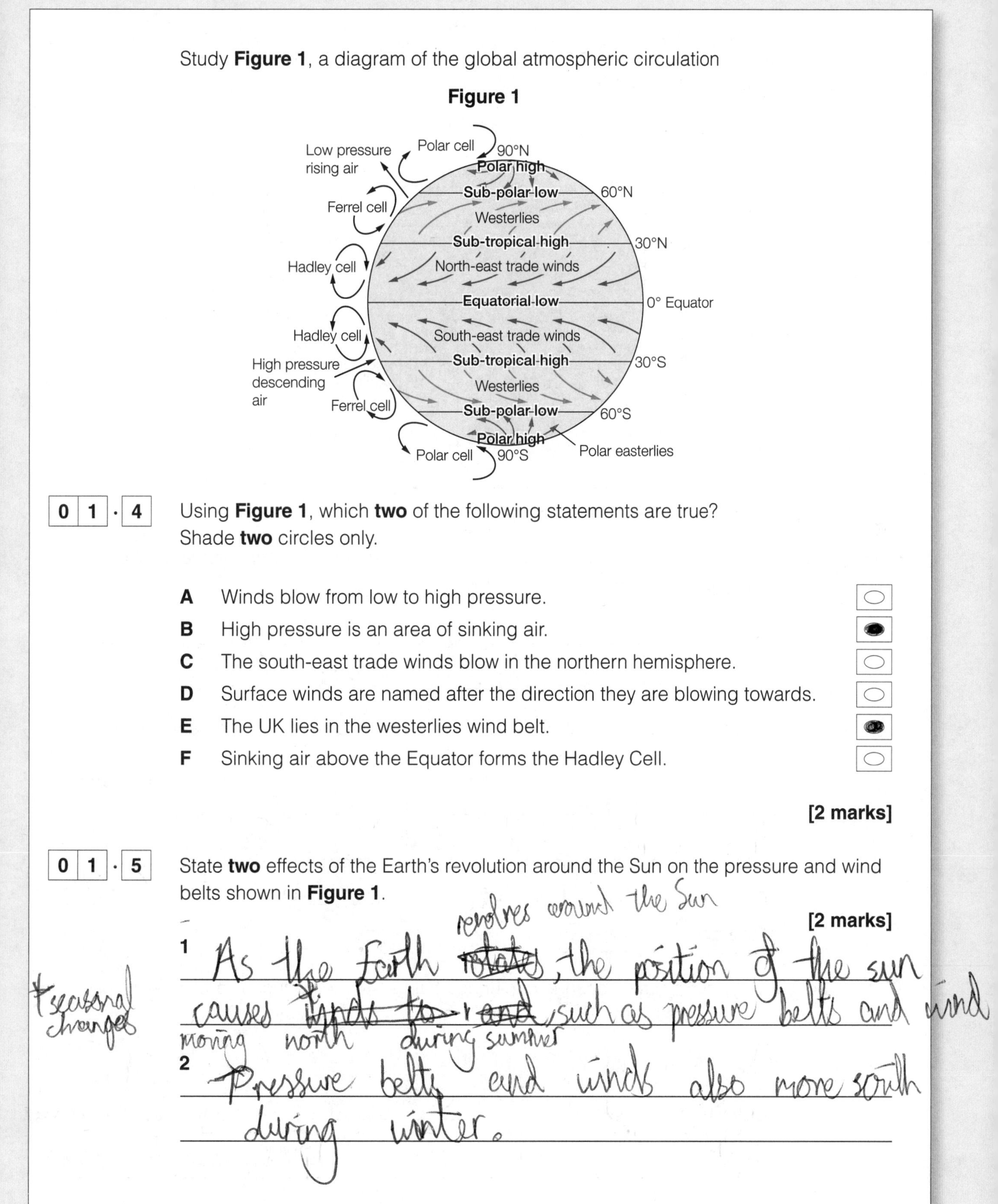

0 1 · 4 Using **Figure 1**, which **two** of the following statements are true?
Shade **two** circles only.

A Winds blow from low to high pressure. ○
B High pressure is an area of sinking air. ●
C The south-east trade winds blow in the northern hemisphere. ○
D Surface winds are named after the direction they are blowing towards. ○
E The UK lies in the westerlies wind belt. ●
F Sinking air above the Equator forms the Hadley Cell. ○

[2 marks]

0 1 · 5 State **two** effects of the Earth's revolution around the Sun on the pressure and wind belts shown in **Figure 1**.

[2 marks]

1 As the Earth ~~rotates~~ revolves around the Sun, the position of the sun causes ~~winds to~~ ~~and~~, such as pressure belts and wind moving north during summer

+ seasonal changes

2 Pressure belts and winds also move south during winter.

Study **Figure 2**, a graph showing the number of tropical storms in the Atlantic between 1850 and 2010.

Figure 2

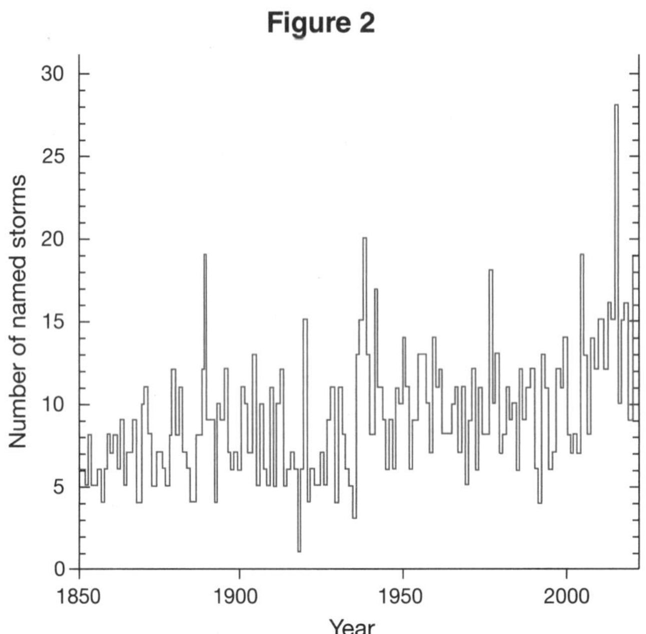

0 1 · 6 What overall change has occurred in the number of storms experienced in the period shown on **Figure 2**?

[1 mark]

Overall the number of storms seem to have increased but there have been fluctuations

0 1 · 7 Explain why most tropical storms develop between 5° and 15° north and south of the Equator.

[2 marks]

As these places have warm oceans (27°C) and there is not enough 'spin' from the rotation of the Earth. This effect is called the Coriolis ~~Effect~~ effect.

0 1 · 8 Explain how climate change can bring changes to the frequency and intensity of tropical storms.

[6 marks]

As more greenhouse gases are released from burning fossil fuels, the greenhouse effect, which warms the earth by trapping radiation, is enhanced. As a consequence, radiation is reflected back at Earth and the ~~seas~~ oceans' average temperatures increase.

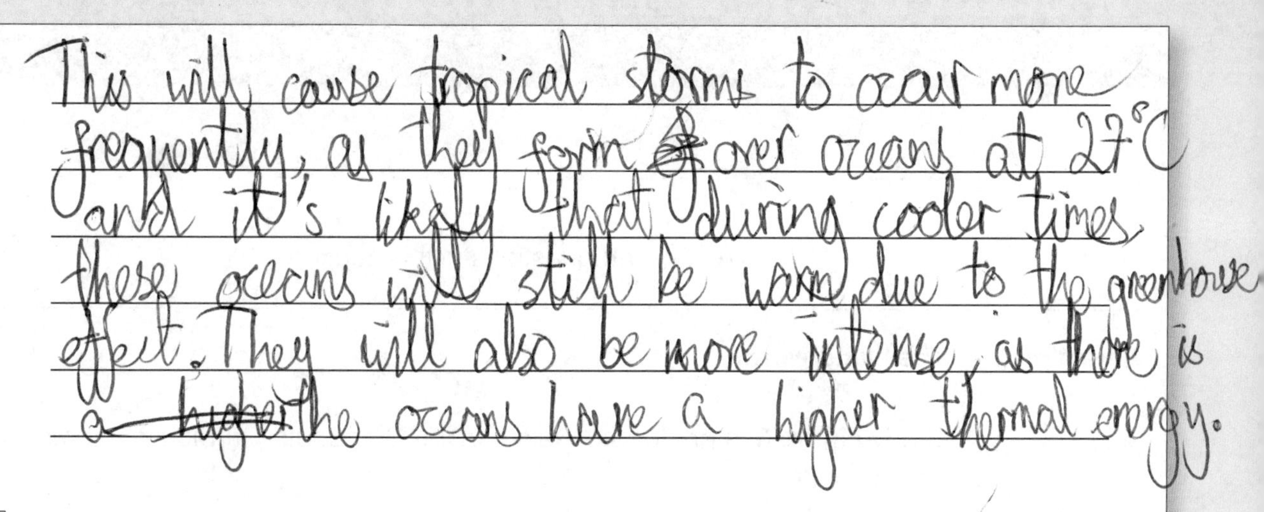
This will cause tropical storms to occur more frequently, as they form over oceans at 27°C and it's likely that during cooler times these oceans will still be warm due to the greenhouse effect. They will also be more intense, as the oceans have a higher thermal energy.

0 1 · 9 Justify why managing climate change needs to be a combination of mitigation and adaption strategies.

[9 marks]
[+3 SPaG marks]

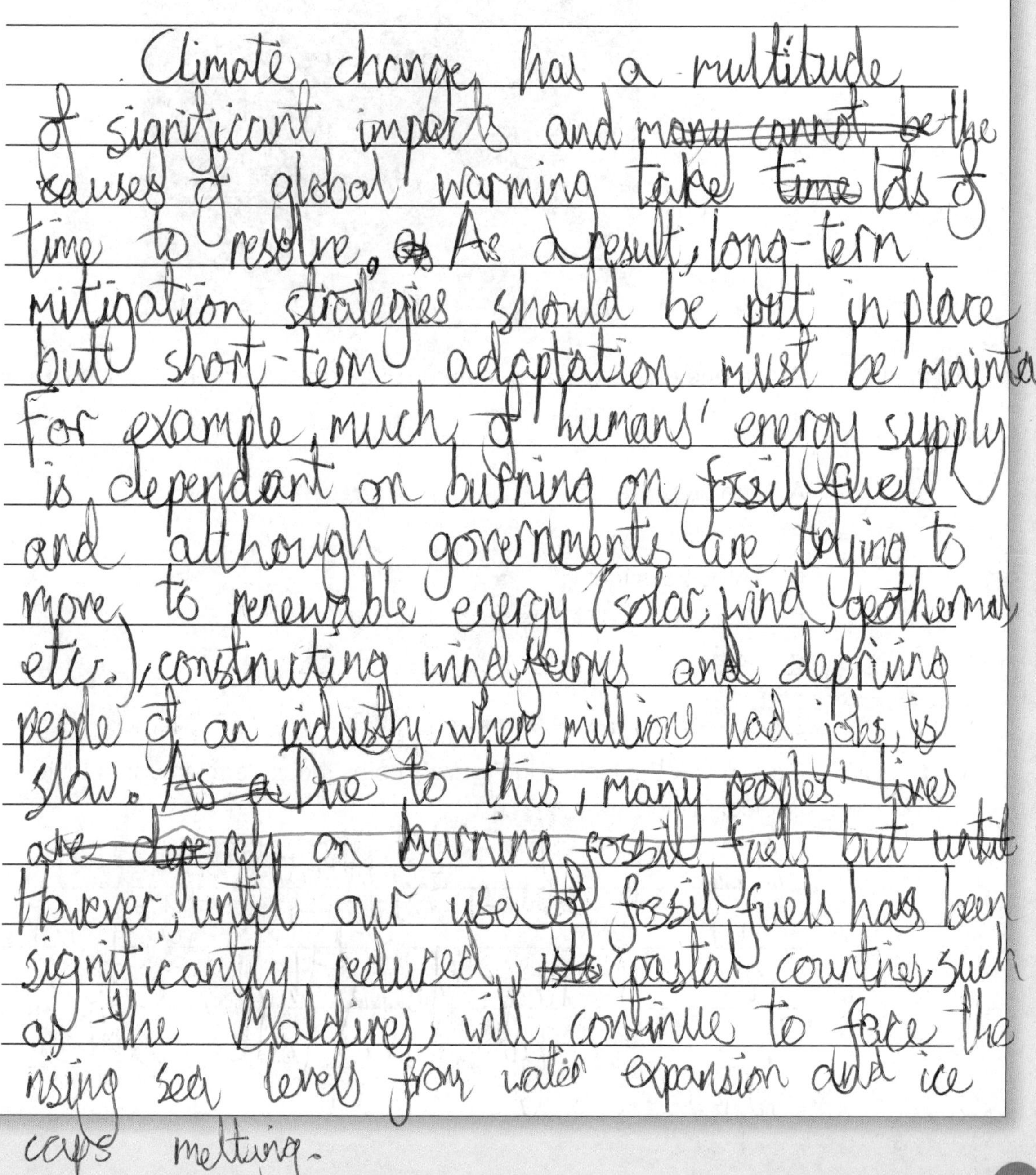
Climate change has a multitude of significant impacts and the causes of global warming take lots of time to resolve. As a result, long-term mitigation strategies should be put in place but short-term adaptation must be maintained. For example, much of humans' energy supply is dependant on burning on fossil fuels and although governments are trying to move to renewable energy (solar, wind, geothermal etc.), constructing wind farms and depriving people of an industry where millions had jobs, is slow. Due to this, many peoples' lives rely on burning fossil fuels. However, until our use of fossil fuels has been significantly reduced, coastal countries such as the Maldives, will continue to face the rising sea levels from water expansion and ice caps melting.

Section B The living world

Answer **all** questions in this section

Question 2 The living world

Study **Figure 3**, a photograph taken in the Cambodian tropical rainforests in south-east Asia.

Figure 3

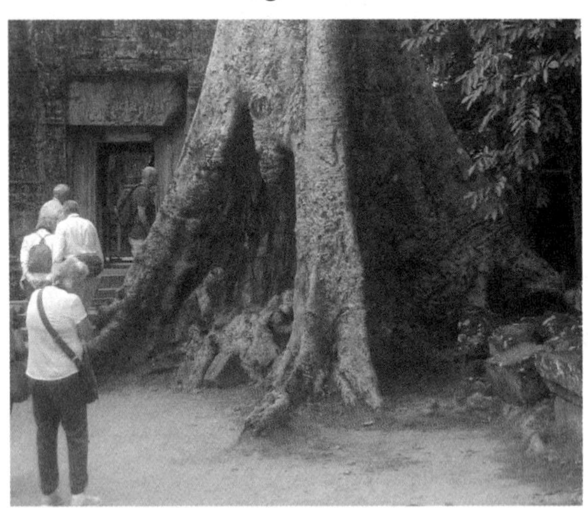

0 2 . 1 Which **one** of the following features of a tropical rainforest is shown in **Figure 3**?
Shade **one** circle only.

A An epiphyte ○
B The canopy ○
C A buttress root ●
D A liana ○
E A drip tip ○

[1 mark]

0 2 . 2 Explain how the feature in **Figure 3** helps the vegetation adapt to the climate of the tropical rainforests.

[2 marks]

1) Due to the significant vegetation, trees often have to compete to reach sunlight and having strong thick roots allows for trees to grow very tall, so they can photosy-nthesise.

Study **Figure 4**, diagram showing the food web in a tropical rainforest.

Figure 4

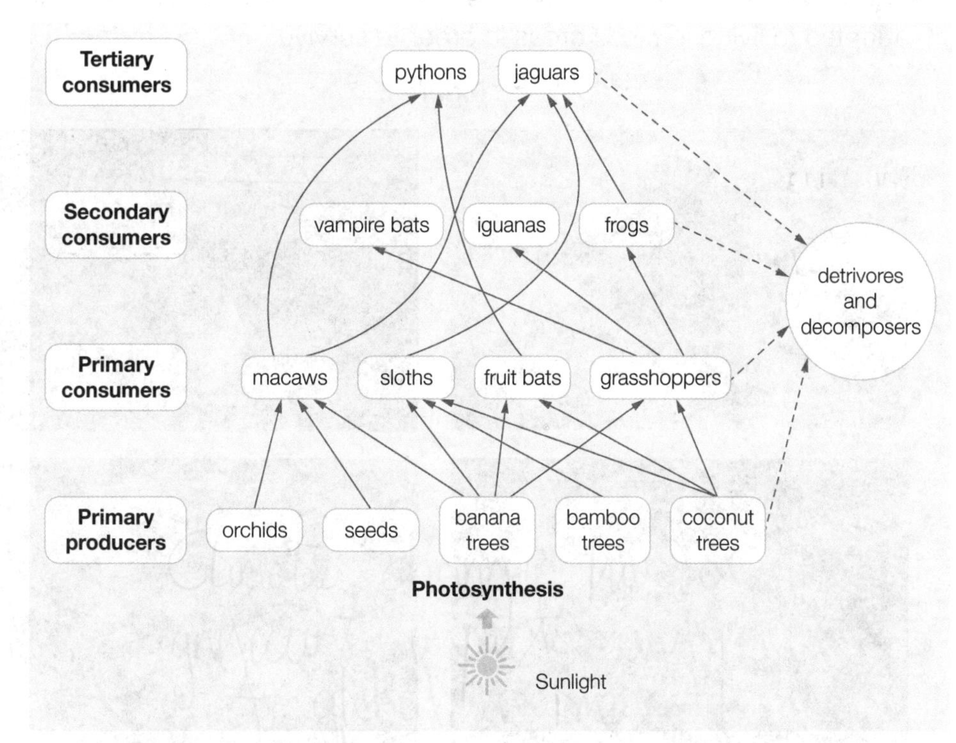

0 2 · 3 Use **Figure 4** to complete the following table by adding **one** example in each blank cell. An example of detrivores and decomposers, fungi, has already been added.

Primary producers	Primary consumers	Secondary consumers	Tertiary consumers	Detrivores
Banana tree	Macaw	Vampire bat	Pythons	fungi
Plants	Herbivores	Carnivores	Top carnivores	Decomposers

[3 marks]

Study **Figure 5** and **Figure 6**.
Figure 5 shows some animals that live in tropical rainforests.
Figure 6 lists some of the ways animals living in the tropical rainforests have adapted to living in these areas in order to survive.

Figure 5

Figure 6

- Camouflage
- Sleeping throughout the day
- Having a very specialist diet
- Living in the forest canopy

0 2 · 4 Use **Figure 5** and **Figure 6** and your own knowledge to explain how animals have adapted to the physical conditions of the tropical rainforests.

[4 marks]

0 2 · 5 Justify why tropical rainforests should be protected.

[6 marks]

Continue your answer on a separate sheet of paper if necessary

0 2 · 6 Assess the importance of the interdependence of the climate, soils and people in **either** a hot desert environment **or** a cold environment.

[9 marks]

Continue your answer on a separate sheet of paper if necessary

Section C Physical landscapes in the UK

Answer **two** questions from the following:
Question 3 (Coasts), Question 4 (Rivers), Question 5 (Glacial)

Question 3 Coastal landscapes in the UK

Study **Figure 7**, a map showing the major physical features of the British Isles.

Figure 7

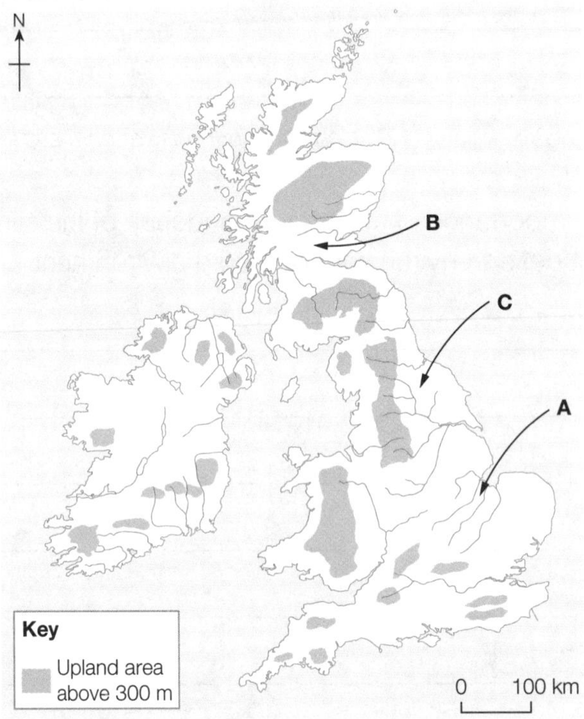

0 3 · 1 Complete the following table by adding the correct letter against each lowland area.

Lowland area	Letter
Central Lowlands of Scotland	
The Fens	
Vale of York	

[2 marks]

Study **Figure 8**, a diagram of a coastal process.

Figure 8

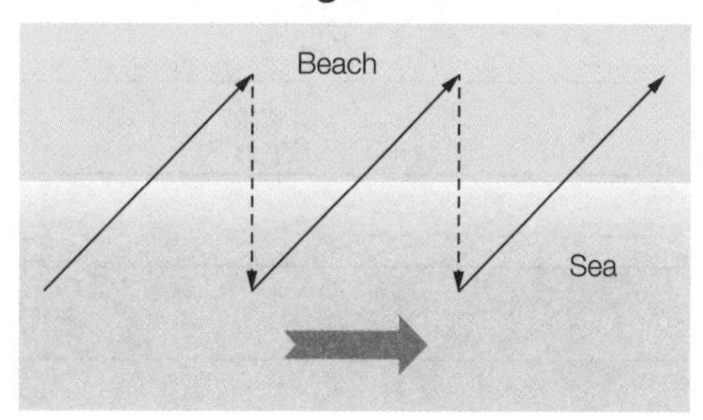

0 3 · 2 What is the name of this coastal process?

[1 mark]

---

0 3 · 3 On a copy of **Figure 8**, label the following:

- direction of coastal process
- backwash
- swash

[2 marks]

Study **Figure 9**, a photograph of a landslide at West Bay in Dorset.

Figure 9

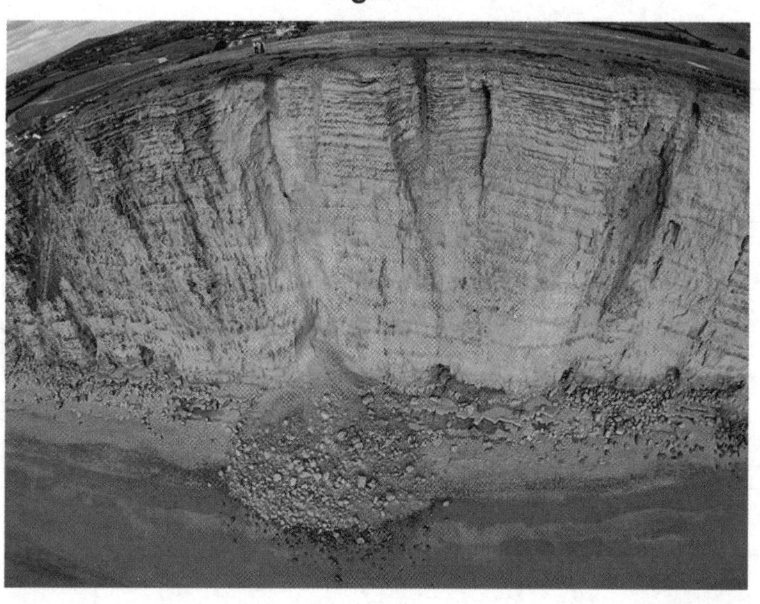

0 3 · 4 Using **Figure 9** and your own knowledge, explain how mass movement can affect the shape of the coastline.

[4 marks]

See over for answer lines

0 3 · 5 Name an example of a coastal management scheme in the UK.

Assess whether the overall benefits outweigh any conflicts that are caused as a result of the scheme.

[6 marks]

Question 4 River landscapes in the UK

Study **Figure 10**, a map showing the major physical features of the British Isles.

Figure 10

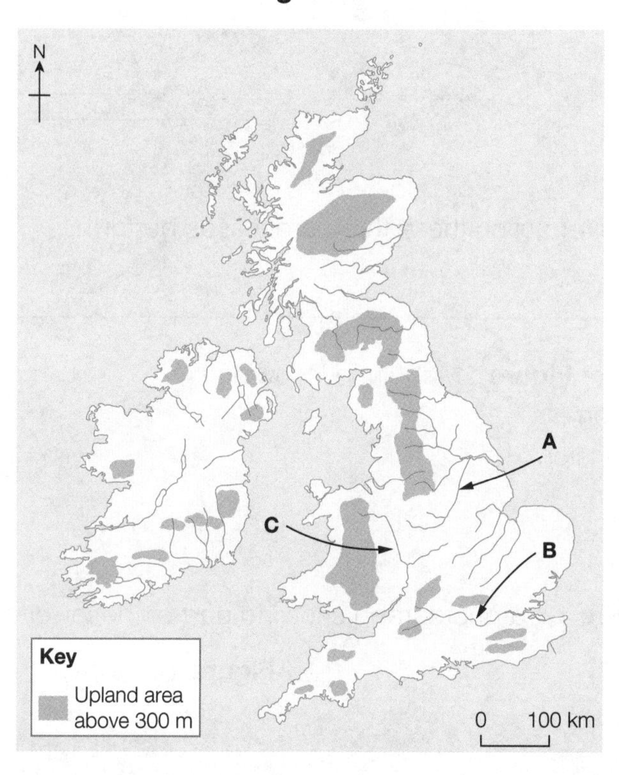

0 4 · 1 Complete the following table by adding the correct letter against each river name.

River	Letter
River Severn	
River Thames	
River Trent	

[2 marks]

Study **Figure 11**, a diagram showing three river processes.

Figure 11

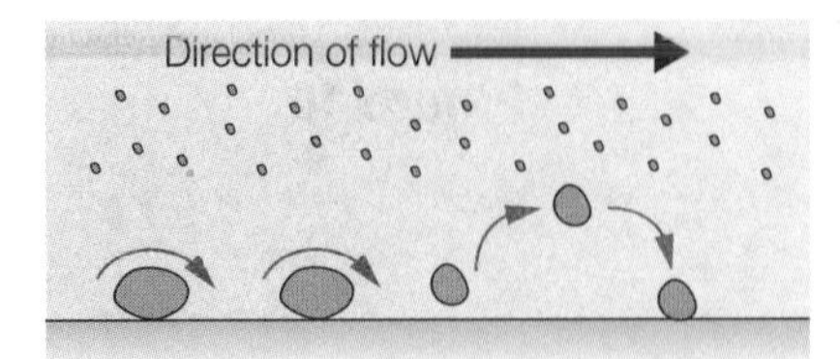

0 4 · 2 Name the river action these three processes perform.

[1 mark]

---

0 4 · 3 On a copy of **Figure 11** label the following:

- saltation
- suspension
- traction

[2 marks]

Study **Figure 12**, a photograph showing part of the lower course of a river.

Figure 12

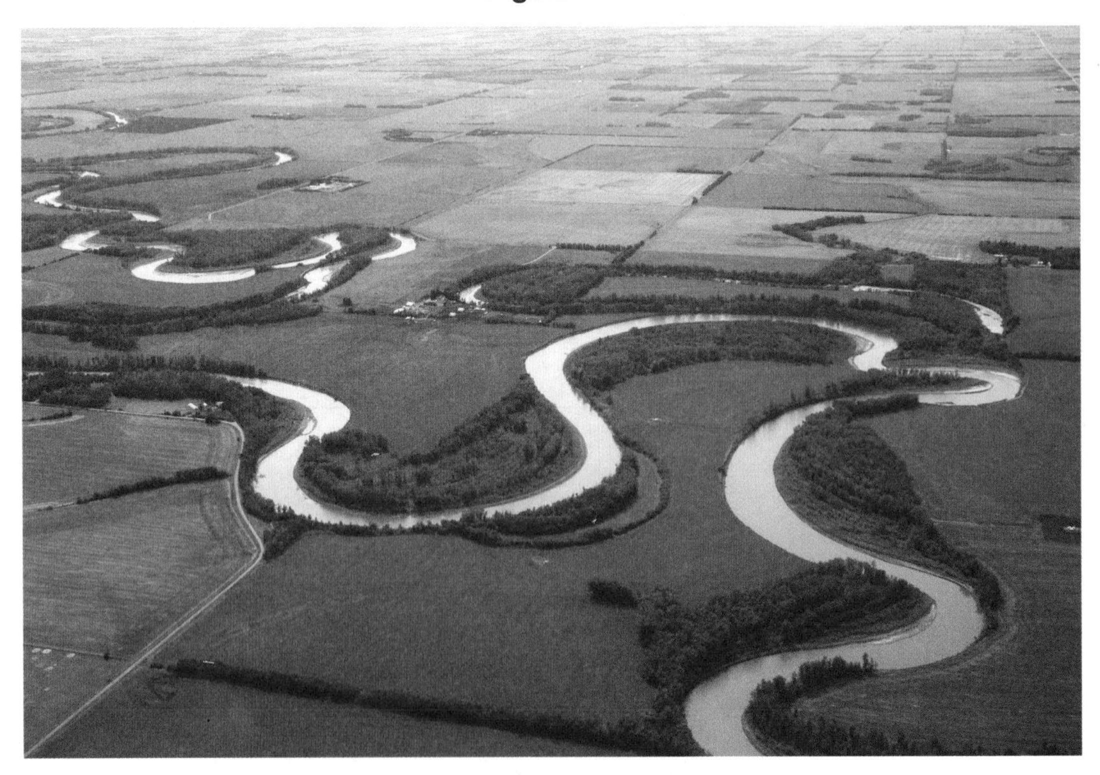

0 4 · 4 Using **Figure 12** and your own knowledge, explain how meanders contribute to the shape of the cross section of a river valley in its lower course.

[4 marks]

0 4 · 5 Name an example of a flood management scheme in the UK.

Assess whether the overall benefits outweigh any environmental issues that are caused as a result of the scheme.

[6 marks]

Question 5 Glacial landscapes in the UK

Study **Figure 13**, a map showing the major physical features of the British Isles.

Figure 13

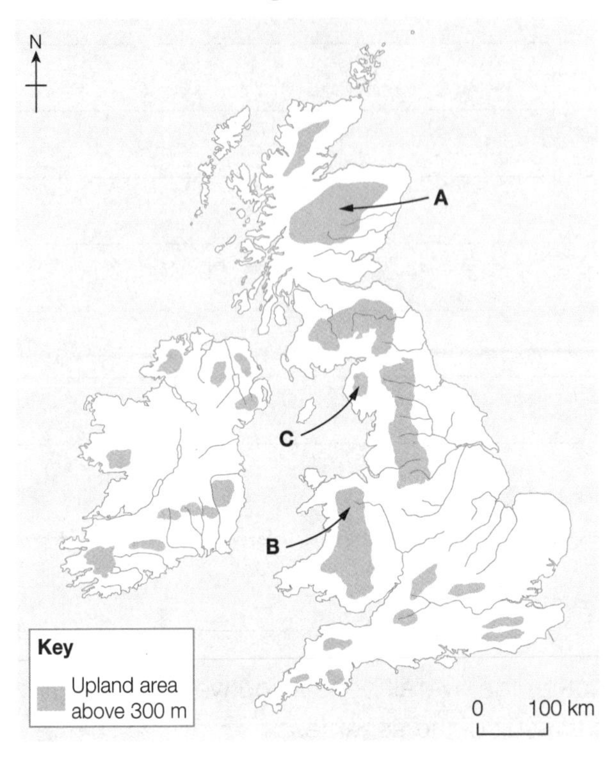

0 5 · 1 Complete the following table by adding the correct letter against each upland area name.

Upland area	Letter
Grampians	
Lake District	
Snowdonia	

[2 marks]

Study **Figure 14**, a diagram showing areas of glacial deposition

Figure 14

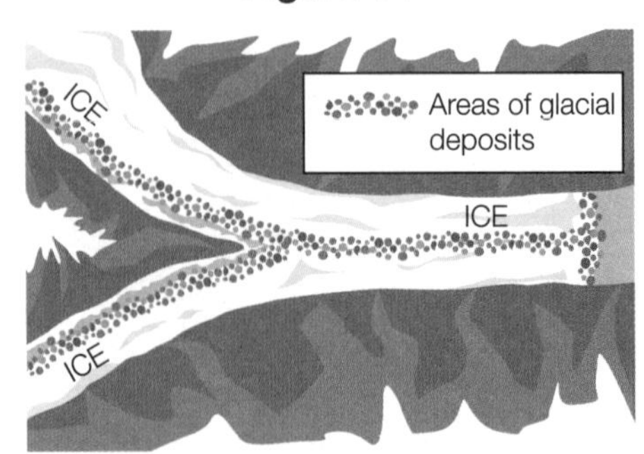

0 5 · 2 What is the name given to glacial deposits?

[1 mark]

____

0 5 · 3 On a copy of **Figure 14** label the following glacial deposits:

- lateral
- medial
- terminal

[2 marks]

Study **Figure 15**, a photograph showing a glaciated highland area.

Figure 15

0 5 · 4 Using **Figure 15** and your own knowledge, explain how glaciation has affected the shape of the landscape.

[4 marks]

____

____

____

____

____

____

____

____

0 5 · 5 Name an example of a glaciated upland area in the UK where tourism is important.

___

Assess whether the overall benefits outweigh any environmental damage that is caused as a result of tourism.

[6 marks]

GCSE 9-1 Geography AQA
Practice Paper

Paper 2 Challenges in the human environment

Time allowed: 1 hour 30 minutes
Total number of marks: 88 (including 3 marks for spelling, punctuation, grammar and specialist terminology (SPaG))

Instructions
Answer **all** questions in Section A and Section B
Answer question 3 and **one** from questions 4, 5 or 6

Section A Urban issues and challenges

Answer **all** questions in this section

Question 1 **Urban issues and challenges**

Study **Figure 1**, showing the population change in a major world city.

Figure 1

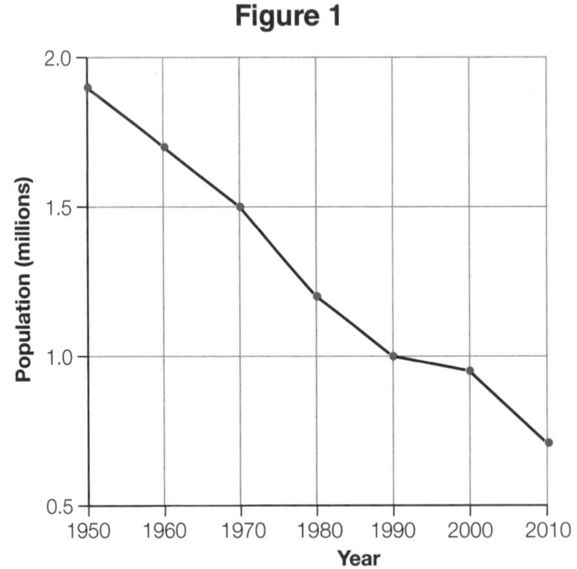

0 1 · 1 What was the population of the city in 1970?

[1 mark]

____

0 1 · 2 What was the approximate change in the population between 1950 and 2010?

[1 mark]

____

0 1 · 3 Suggest in which type of country this city is likely to be situated.

Shade the correct circle.

- ○ High-income country (HIC)
- ○ Low-income country (LIC)

[1 mark]

0 1 · 4 Give **two** reasons for your answer.

Reason 1

____

____

Reason 2

____

____

[2 marks]

Study **Figure 2**, which gives information about Dharavi, a squatter settlement in the Indian city of Mumbai.

Figure 2

People	
Population of Dharavi	Estimated 800 000–1 million
Area	2.39 km^2 (the size of London's Hyde Park)
Population density	At least 330 000 people per km^2
No of homes in Dharavi	60 000
People per home	Between 13 and 17
Average size of home	10 m^2 (equivalent to a medium-sized bedroom)

**anaemia is a lack of iron leading to tiredness*
*** gastro-enteritis symptoms are diarrhoea and vomiting*

Hygiene and health	
No of individual toilets in Dharavi	1440
People per individual toilet	625
% of women suffering from anaemia*	75%
% of women with malnutrition	50%
% of women with recurrent gastro-enteritis**	50%
Most common causes of death	Malnutrition, diarrhoea, dehydration, typhoid
Education	
Literacy rate in Dharavi	69% (Mumbai average is 91%)

0 1 · 5 With the help of **Figure 2**, explain why urban growth in LICs and NEEs often leads to serious challenges for the city.

[4 marks]

____

____

____

____

____

____

____

____

0 1 · 6 Consider the attempts of a city in an LIC or an NEE to provide sufficient health and education services for its inhabitants.

[6 marks]

Study **Figure 3**, a map showing the clustering of Asian Indian British residents in London.

Figure 3

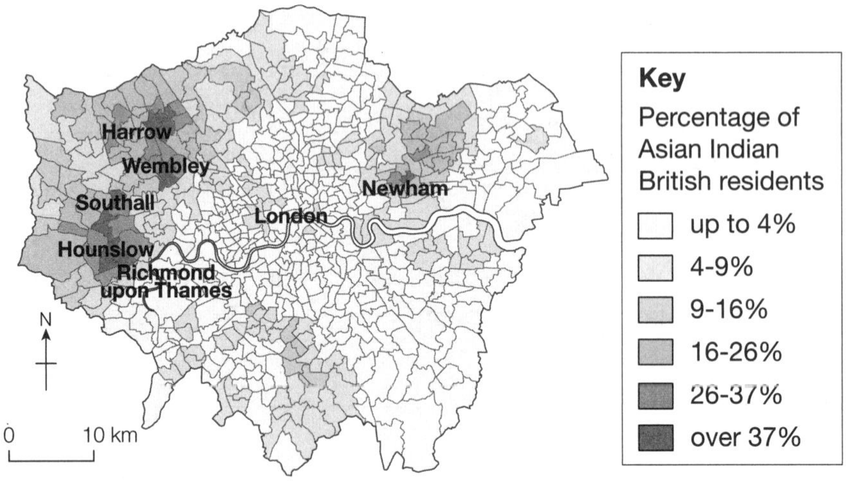

Harrow
Wembley
Southall
Hounslow
Richmond upon Thames
London
Newham
N
0 10 km

Key
Percentage of Asian Indian British residents
up to 4%
4-9%
9-16%
16-26%
26-37%
over 37%

0 1 · 7 Using **Figure 3** describe the distribution of Asian Indian British residents in London.

[2 marks]

0 1 · 8 Explain how migration can affect the character of the city.

[4 marks]

0 1 · 9 Use a case study of a major UK city to assess the extent to which urban change has created social and economic opportunities for the city.

[9 marks]

[+3 SPaG marks]

Section B The changing economic world

Answer **all** questions in this section

Question 2 The changing economic world

Study **Figure 4**, a partly completed diagram of the five stages of the Demographic Transition Model.

Figure 4

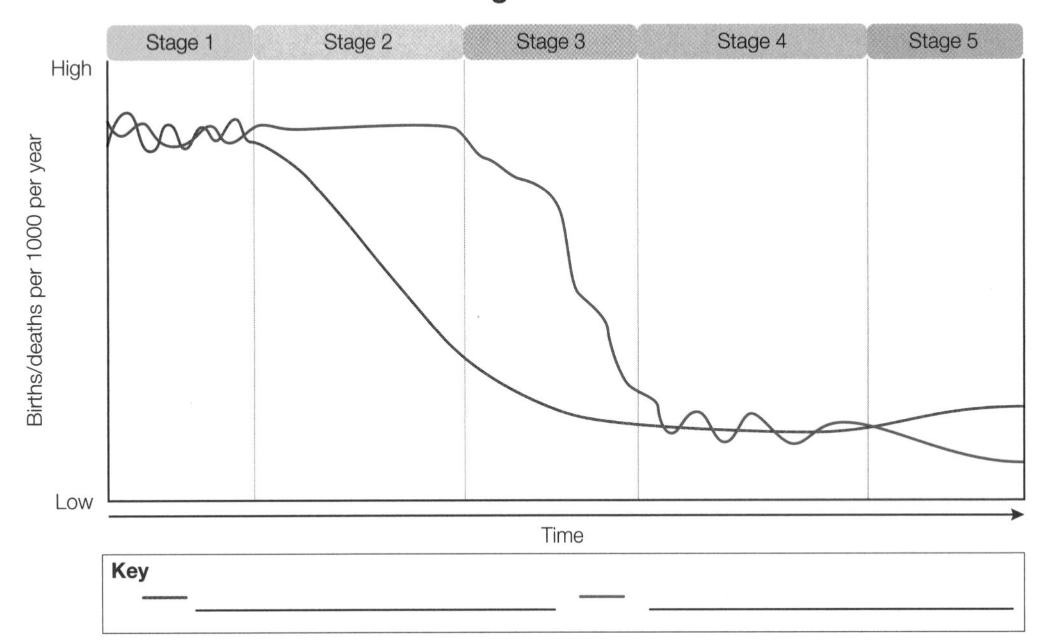

0 2 · 1 On **Figure 4**, complete the key by adding 'Birth rate' and 'Death rate' to the correct line symbol.

[1 mark]

0 2 · 2 On **Figure 4**, shade the area where there is greatest population increase.

[1 mark]

0 2 · 3 Complete the table to show which is the most likely stage that the following types of country have reached.

Type of country	Stage
HIC	
LIC	
NEE	

[3 marks]

0 2 . 4 Explain how **physical** factors can cause uneven development.

[4 marks]

0 2 . 5 Name the LIC or NEE you studied as your case study.

Name of country ______________________

Describe its location.

[2 marks]

0 2 . 6 Explain why the country is important **either** regionally **or** internationally.

[4 marks]

Study **Figure 5**, two photographs of contrasting examples of economic development in India.

Figure 5

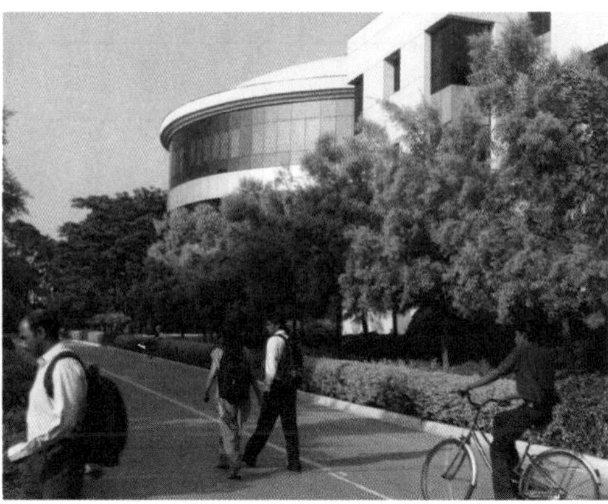

0 2 . 7 Assess the environmental impacts of economic development using evidence from **Figure 5** and your own knowledge of an LIC or NEE.

[6 marks]

Study **Figure 6**, which gives details of the north–south divide in Great Britain.

Figure 6

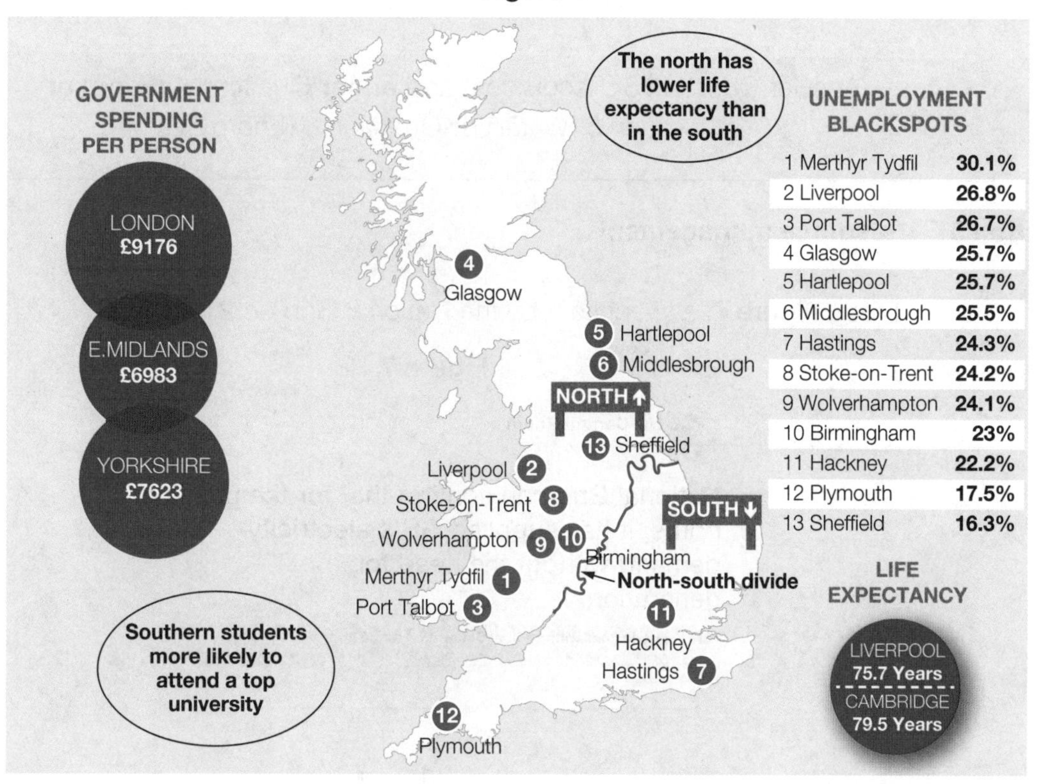

0 2 · 8 With the help of **Figure 6**, evaluate the strategies that attempt to resolve the differences between the different parts of the country.

[9 marks]

Continue your answer on a separate sheet of paper if necessary

Section C The challenge of resource management

Answer Question 3 (Resources) and **either** Question 4 (Food) **or** Question 5 (Water) **or** Question 6 (Energy)

Question 3 Resource management

Study **Figure 7**, a tweet sent by the National Grid on 21 April 2017.

Figure 7

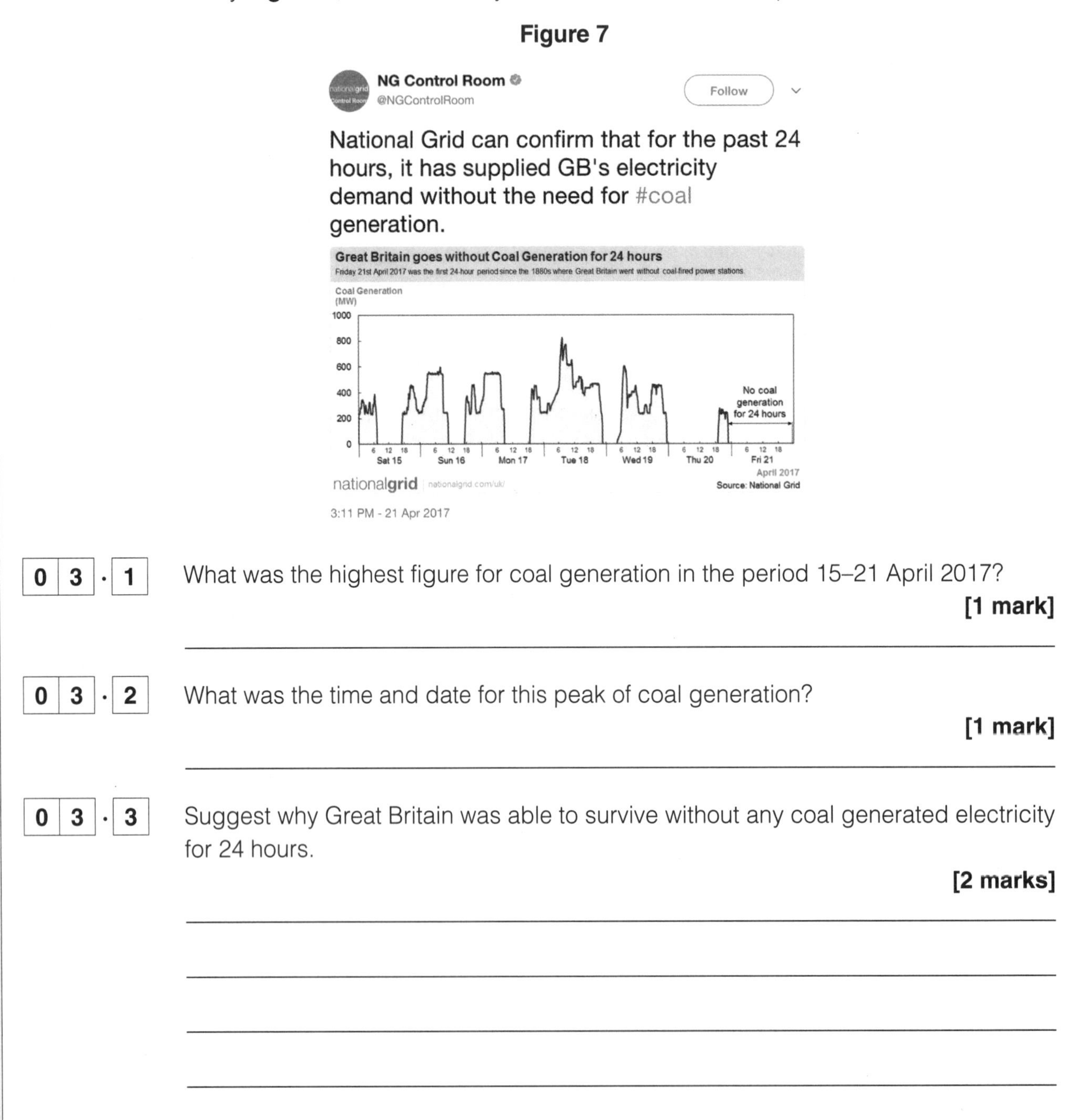

0 3 · 1 What was the highest figure for coal generation in the period 15–21 April 2017?

[1 mark]

____

0 3 · 2 What was the time and date for this peak of coal generation?

[1 mark]

____

0 3 · 3 Suggest why Great Britain was able to survive without any coal generated electricity for 24 hours.

[2 marks]

____

____

____

____

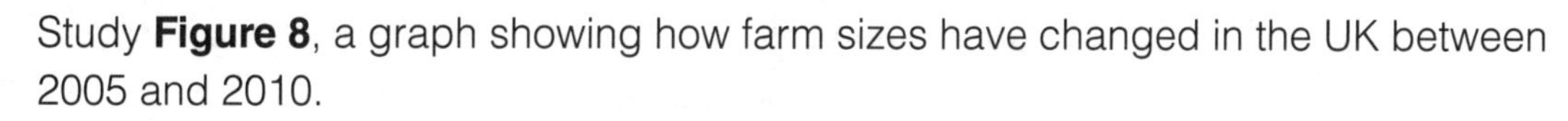

Study **Figure 8**, a graph showing how farm sizes have changed in the UK between 2005 and 2010.

Figure 8

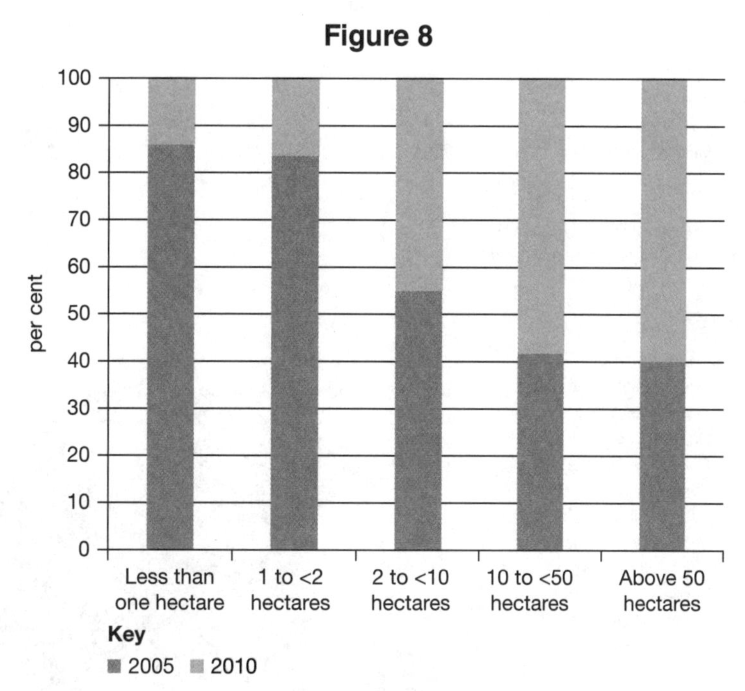

0 3 · 4 What percentage of the farms were between 2 and less than 10 hectares in 2010?

[1 mark]

____

0 3 · 5 State what **Figure 8** shows about how the size of UK farms changed between 2005 and 2010.

[1 mark]

____

____

0 3 · 6 Suggest how the change in the size of UK farms might affect the provision of food in the UK.

[2 marks]

____

____

____

____

Study **Figure 9**, which shows areas of water stress in England.

Figure 9

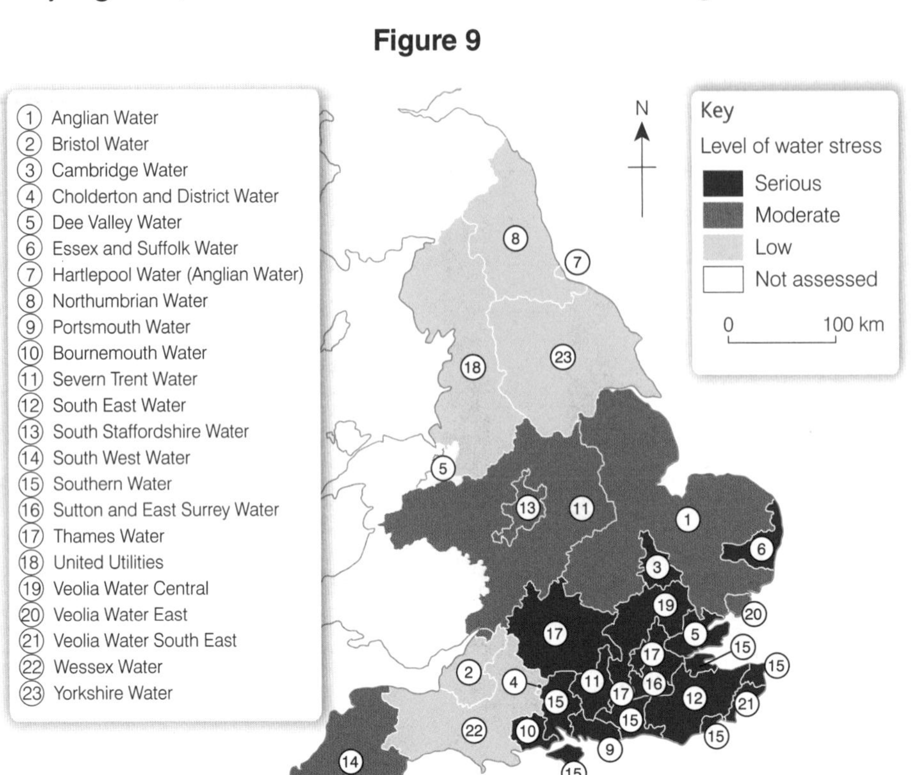

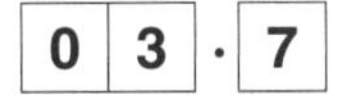
0 3 · 7

With the help of **Figure 9**, discuss how water transfer may be needed to maintain supplies across England.

[6 marks]

Answer **either** Question 4 (Food) **or** Question 5 (Water) **or** Question 6 (Energy).

Question 4 Food

Study **Figure 10**, which gives information about the country of South Sudan in Africa.

Figure 10

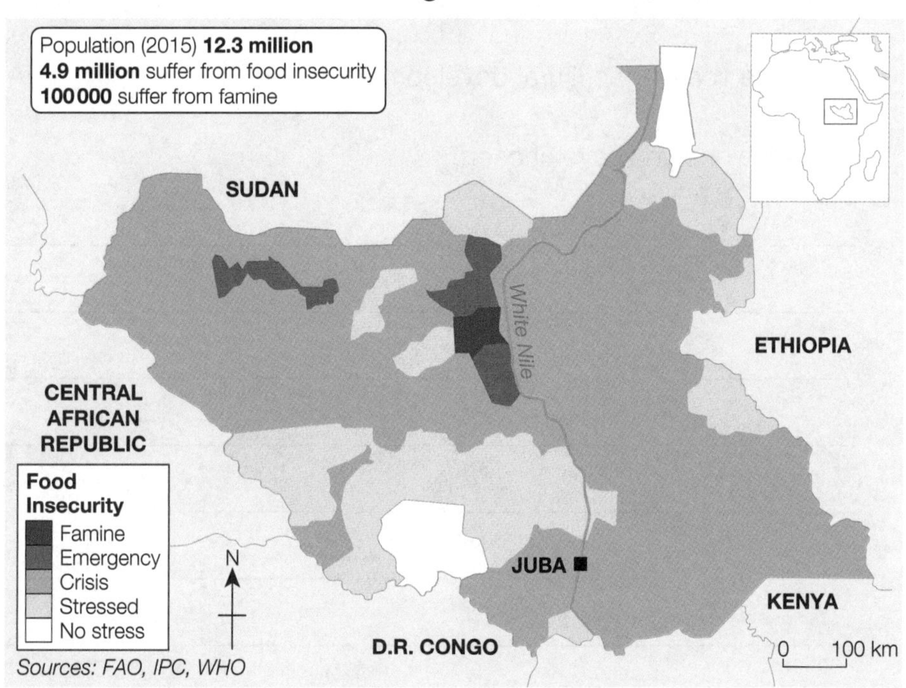

0 4 · 1 What is meant by food insecurity?

[1 mark]

____

____

0 4 · 2 Give **two** pieces of evidence from **Figure 9** that shows the South Sudan suffers from food insecurity.

[2 marks]

1 ____

2 ____

0 4 · 3 Name **two** impacts of food insecurity on a country.

[2 marks]

1 ____

2 ____

0 4 · 4 Choose an example of **either** a large-scale agricultural development **or** a local scheme in an LIC or a NEE which aims to increase the supply of food.

For the example chosen, discuss the extent to which it has been able to increase the supply of food.

Name of example ______________________________

A large scale agricultural development   A local scheme

Circle the one you have chosen.

[6 marks]

Question 5 Water

Study **Figure 11**, which gives information about the Nile Basin.

Figure 11

While Egypt is entirely dependent on the Nile for its water supply and regards any possible reduction as an issue of national security, some of the world's poorest countries see the river as a vital source for national development.

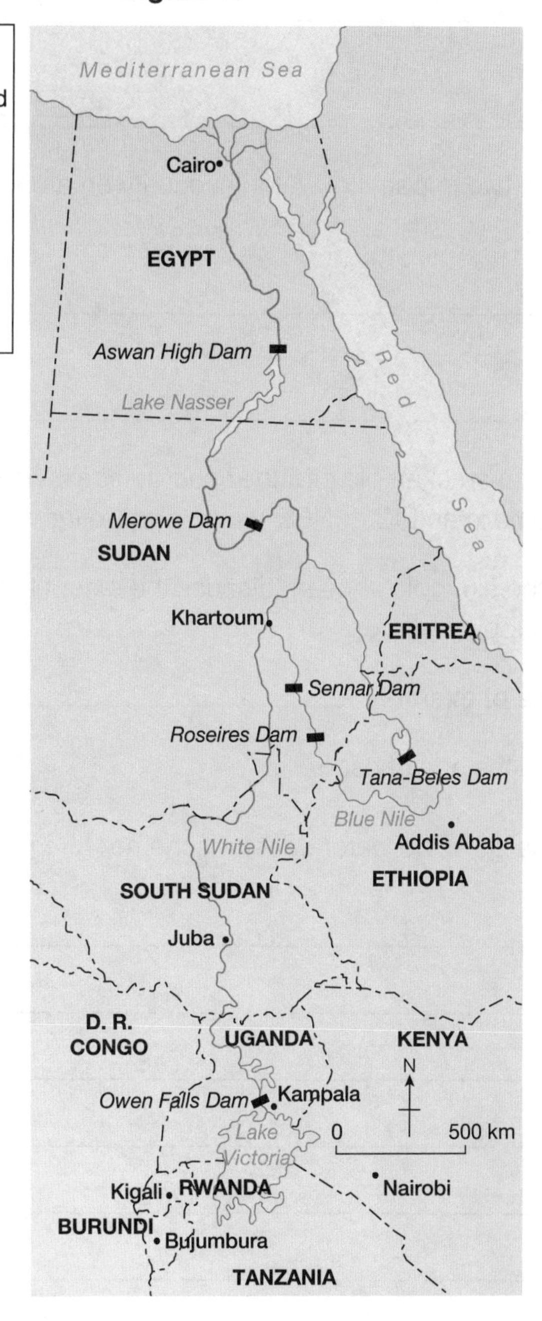

0 5 · 1 What is meant by water insecurity?

[1 mark]

____

____

0 5 · 2 Give **two** pieces of evidence from **Figure 11** that show countries in the Nile Basin could suffer from water insecurity.

[2 marks]

1 ______________________________

2 ______________________________

0 5 · 3 Name **two** impacts of water insecurity on a country.

[2 marks]

1 ______________________________

2 ______________________________

0 5 · 4 Choose an example of **either** a large scale water transfer scheme **or** a local scheme in an LIC or NEE, which aims to increase the supply of water.

For the example chosen, discuss the extent to which it has been able to increase the supply of water.

Name of example ______________________________

Circle the one chosen.

A large scale water transfer scheme   A local scheme

[6 marks]

____

____

____

____

____

____

____

____

____

____

____

Question 6 Energy

Study **Figure 12**, which gives information about the gas supplies in Europe.

Figure 12

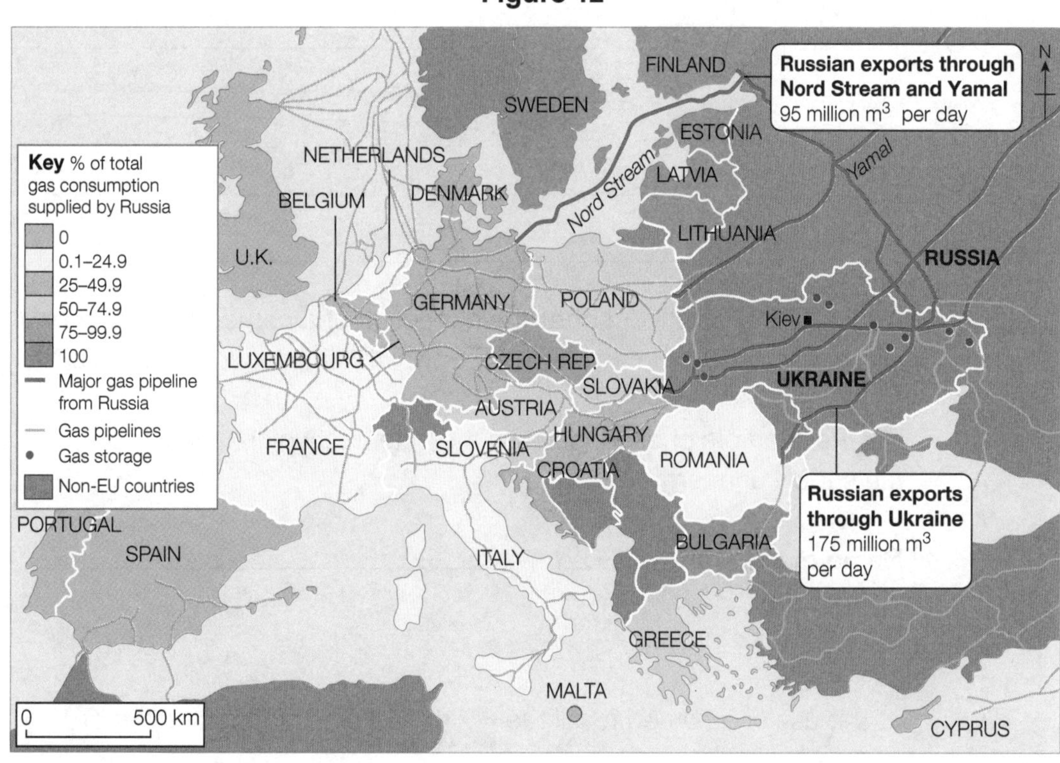

0 6 · 1 What is meant by energy insecurity?

[1 mark]

---

0 6 · 2 Give **two** pieces of evidence from **Figure 12** that shows EU countries could suffer from energy insecurity.

[2 marks]

1

2

---

0 6 · 3 Name **two** impacts of energy insecurity on a country.

[2 marks]

1

2

0 6 · 4 Choose an example of **either** the extraction of a named fossil fuel **or** a local renewable scheme in an LIC or NEE, which aims to increase the supply of energy.

For the example chosen, discuss the extent to which it has been able to increase the supply of energy.

Name of example ______________________

Circle the one chosen.

The use of a named fossil fuel   A local renewable energy scheme

[6 marks]

GCSE 9-1 Geography AQA
Practice Paper

Paper 3 Geographical applications

Time allowed: 1 hour 15 minutes
Total number of marks: 76 (including 6 marks for spelling, punctuation, grammar and specialist terminology (SPaG))

Instructions
Answer **all** questions
Use a clean copy of the pre-release resources booklet

Section A Issue evaluation

Answer **all** questions in this section.

Study **Figure 1** in the resources booklet, 'The tropical rainforest global ecosystem or biome'.

0 1 . 1 Describe the annual rainfall pattern of Belem.

[2 marks]

0 1 . 2 Why is it possible to describe Belem as **either** having one season all year around **or** having two climatic seasons?

[2 marks]

0 1 . 3 How can climate change interrupt the nutrient cycle of the tropical rainforests?

[2 marks]

0 1 · 4 What percentage of the deforestation in Amazonia is caused by cattle ranching?

Shade **one** circle only

[1 mark]

A 40 ○

B 50 ○

C 60 ○

D 70 ○

0 1 · 5 With the help of **Figure 1**, assess the significance of physical and human factors as being responsible for changing the characteristics of the tropical rainforest biome.

[6 marks]

Study **Figure 2** in the resources booklet, 'Biofuels in Indonesia'.

0 2 · 1 Suggest why Indonesia's rainforests have such high levels of diversity.

[4 marks]

0 2 · 2 Explain why the Indonesian Government is so keen to exploit the country's rainforests.

[6 marks]

Study **Figure 3** in the resources booklet, 'The threat to the Indonesian rainforests'.

0 3 · 1 What is the link between the increase in palm oil production and the decline in Indonesian rainforests?

[2 marks]

The following **three** options have been suggested for how Indonesia could manage the country's rainforests in the future.

Option 1	Continue to expand palm oil production to help Indonesia's economy grow as quickly as possible.
Option 2	Slow further expansion of oil palm plantations by taxing palm oil production and monitoring and policing remaining rainforests.
Option 3	Ban palm oil production and create forest reserves, which can only be used for small-scale farming and other sustainable land uses.

0 3 · 2 Which of the **three** options do you think will benefit Indonesia without causing irreparable damage to the country's rainforests?

Use evidence from the resources booklet and your own understanding to explain why you have reached this decision.

[9 marks]
[+ 3 SPaG marks]

Chosen option ____________________

Continue your answer on a separate sheet of paper

Section B Fieldwork

Answer **all** questions

As part of their investigation of the rural geography in part of the county of Dorset, GCSE students were given a set of secondary data on the facilities in a number of villages in the county.

The data is shown in **Figure 4**.

Figure 4

Village	Population	Church/chapel	Village hall	Primary school	Post office & shop	Food shop	Garage	Bank	Doctor	Pub	Playing field	Cash point	Mobile library	Bus service, daily (D) or weekly (W)
Corfe Castle	980	Yes	1	0	1	2	1	0	1	4	1	1	0	D
East Lulworth	170	Yes	0	0	0	0	0	0	0	1	1	0	1	D
Harmans Cross	340	Yes	1	0	1	0	1	0	0	0	0	1	1	D
Kingston	100	Yes	0	0	0	0	0	0	0	1	0	0	0	W
Langton Maltravers	910	Yes	1	1	1	0	0	0	0	2	1	0	1	D
Ridge	290	No	0	0	0	0	0	0	0	0	0	0	1	W
Steeple	30	Yes	0	0	0	0	0	0	0	0	0	0	0	W
Studland	540	Yes	1	0	1	0	0	0	0	1	1	0	1	D
West Lulworth	770	Yes	1	1	1	0	0	0	0	2	1	1	1	D
Wool	1970	Yes	0	2	1	2	2	0	1	2	1	1	0	D
Worth Maltravers	240	Yes	1	0	1	0	0	0	0	1	0	0	0	D

Date of Information: 2006

0 4 · 1 Why is this information an example of secondary data?

[1 mark]

____

____

0 4 · 2 What is the modal value for the number of pubs in this survey?

[1 mark]

____

____

0 4 · 3 Give **two** limitations of this data being used as a basis for a conclusion.

[2 marks]

0 4 · 4 The students were told to collect some primary data using a sampling technique.

Suggest **two** possible sampling techniques they could have used.

[2 marks]

0 4 · 5 Choose **one** of these sampling techniques

Give **one** advantage and **one** disadvantage of your chosen technique.

[2 marks]

Advantage:

Disadvantage:

Study **Figure 5**, a photograph of part of the coast of Northern Ireland, and **Figure 6**, a photograph of the seafront at Brighton in south-east England.

Figure 5

Figure 6

0 4 · 6 For **one** of the areas shown in **Figures 5** and **6**, suggest **two** data collection techniques that could be used to carry out a geographical fieldwork.

[2 marks]

Figure number chosen: ______________________________

Data technique 1 ______________________________

____

Data technique 2 ______________________________

____

Choose *either* **Figure 5** *or* **Figure 6**.

0 4 · 7 Annotate your chosen photograph to show some physical and/or human geographical features of the area.

[4 marks]

____

____

____

____

____

____

____

____

0 4 · 8 Explain why it would be necessary to carry out a risk assessment before carrying out geographical fieldwork in the area you have chosen.

[2 marks]

____

____

____

____

0 5 · 1 State the title of your fieldwork enquiry in which **physical** geography data were collected.

Title of fieldwork enquiry ________________________________

Justify **one** of the data presentation techniques used.

[2 marks]

____

____

____

____

0 5 · 2 Evaluate how the timing of your data collection could have affected the validity of your results.

[3 marks]

____

____

____

____

____

____

0 5 · 3 State the title of your fieldwork enquiry in which **human** geography data were collected.

Title of fieldwork enquiry ________________________________

Explain how your data collection for this enquiry could have been improved.

[6 marks]

____

____

____

____

____

____

____

0 5 · 4 Explain how the interpretation of the data you collected for **either** your human geography enquiry **or** your physical geography enquiry helped your geographical understanding of the topic you investigated.

[9 marks]
[+ 3 SPaG marks]

GCSE 9-1 Geography AQA
Practice Paper

Paper 1 Living with the physical environment

Time allowed: 1 hour 30 minutes
Total number of marks: 88 (including 3 marks for spelling, punctuation, grammar and specialist terminology (SPaG))

Instructions

Answer **all** questions in Section A and Section B
Answer **two** questions in Section C

Section A The challenge of natural hazards

Answer **all** questions in this section.

Question 1 The challenge of natural hazards

Study the plate boundary shown in **Figure 1**.

Figure 1

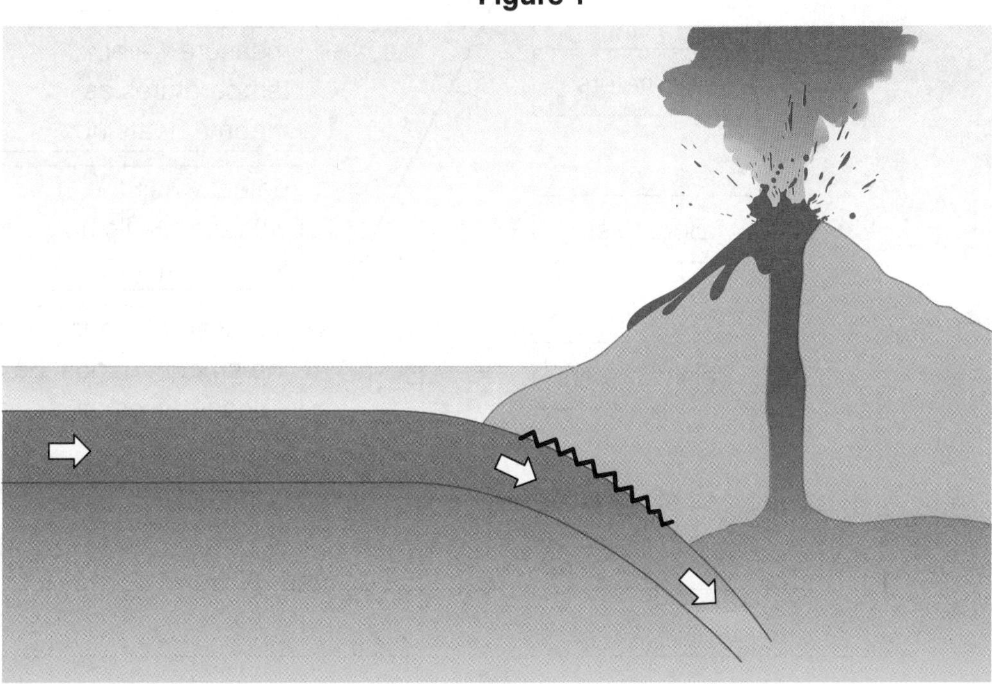

0 1 · 1 Which of the following plate boundaries are shown in **Figure 1**?

Tick the correct answer

[1 mark]

Boundary	Tick
Conservative	
Constructive	
Destructive	

0 1 · 2 Name and label the following features on **Figure 1**.

[5 marks]

- Magma source
- Oceanic plate
- Subduction zone
- Trench
- Volcano

0 1 · 3 Complete the following diagram by connecting the name of the type of volcanic monitoring to the box which shows how it works.

One has been done for you.

[2 marks]

Figure 2

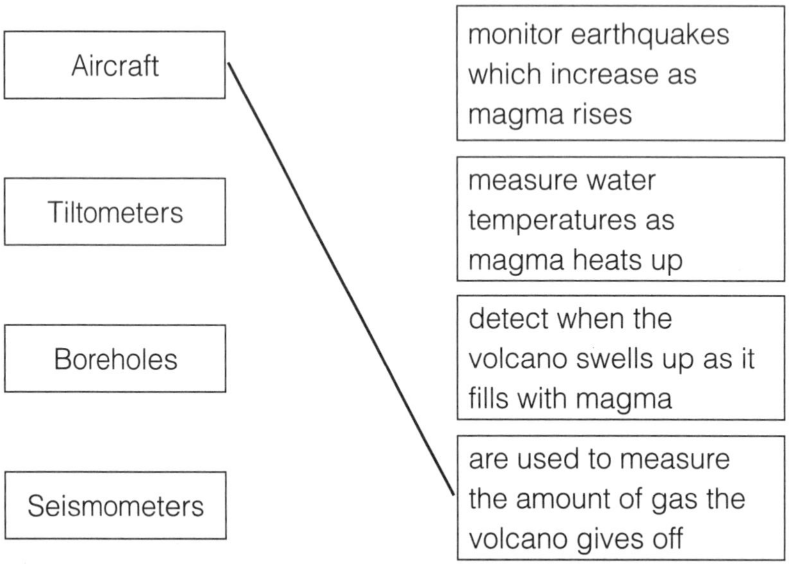

Aircraft	monitor earthquakes which increase as magma rises
Tiltometers	measure water temperatures as magma heats up
Boreholes	detect when the volcano swells up as it fills with magma
Seismometers	are used to measure the amount of gas the volcano gives off

0 1 · 4 Name **two** types of extreme weather experienced in the UK.

[2 marks]

1 ______________________________

2 ______________________________

0 1 · 5 Explain why it is very unlikely that the UK would experience hurricanes.

[4 marks]

____

____

____

____

____

____

____

____

0 1 · 6 Using the example of a recent extreme weather event in the UK, assess whether the socio-economic effects were more important than the environmental effects.

[9 marks] [+ 3 SPaG marks]

0 1 · 7 State **one** effect of global climate change.

[1 mark]

Study **Figure 3**, photographs which show some causes of global climate change.

Figure 3

0 1 · 8 With the aid of **Figure 3** and your own knowledge, justify that climate change is not just the result of human actions.

[6 marks]

Section B The living world

Answer **either** question 02.1 (Hot deserts) **or** question 02.2 (Cold environments).
Then answer **all** remaining questions in this section

Question 2 **The living world**

Answer **either** question 02.1 (Hot deserts) **or** question 02.2 (Cold environments).

0 2 · 1 Study **Figure 4**, a photograph showing a hot desert landscape.

Figure 4

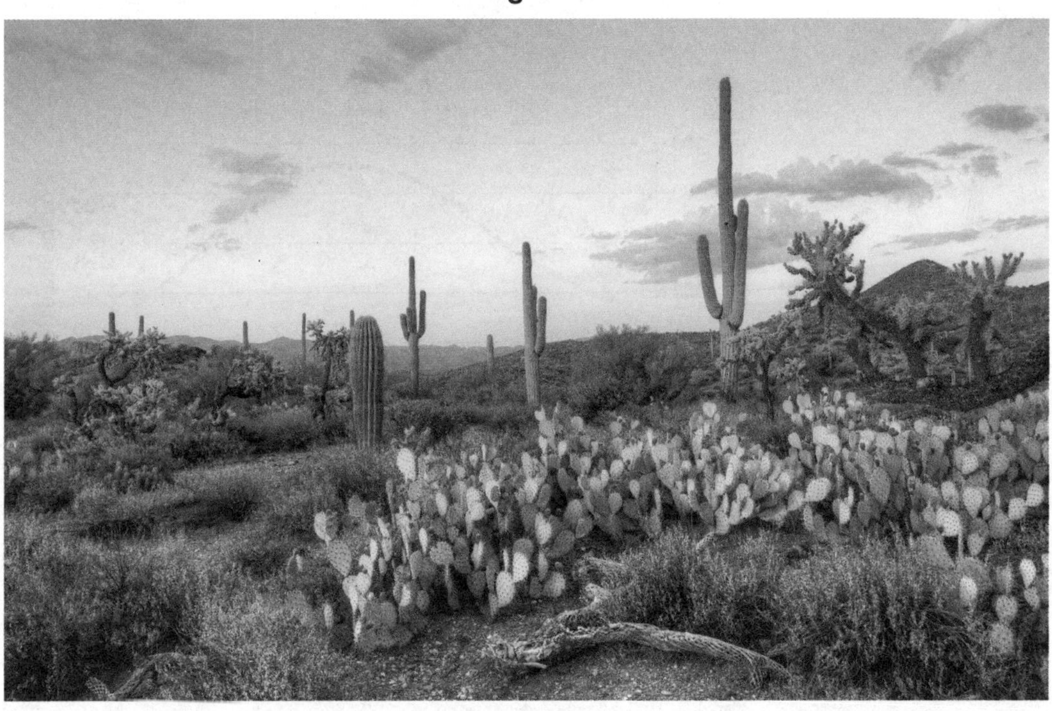

a) Give **three** facts about the vegetation of the hot deserts shown in **Figure 4**.

[3 marks]

____

____

____

____

____

____

Study the three climate graphs in **Figure 5**.

Figure 5

A

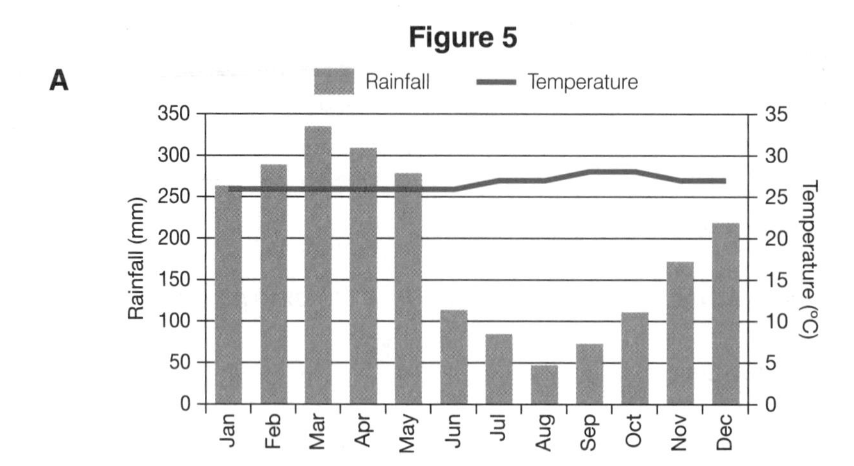

B

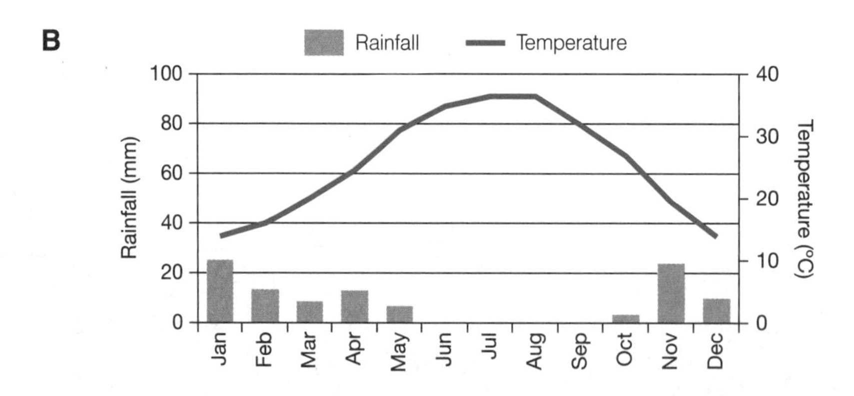

C

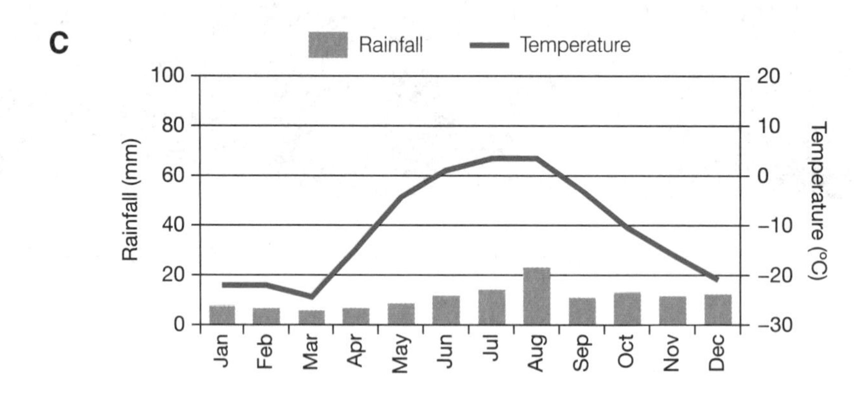

b) Which of the graphs **A, B** or **C** in **Figure 5** represents the climate of a hot desert environment?

Tick the box showing the correct answer.

[1 mark]

Graph	Tick
A	
B	
C	

c) State **one** reason for your choice.

[1 mark]

d) Explain how the vegetation of hot deserts is adapted to the climate.

[6 marks]

Do not answer this question if you have answered question 2.1

0 2 · 2 Study **Figure 6**, a photograph showing a tundra landscape.

Figure 6

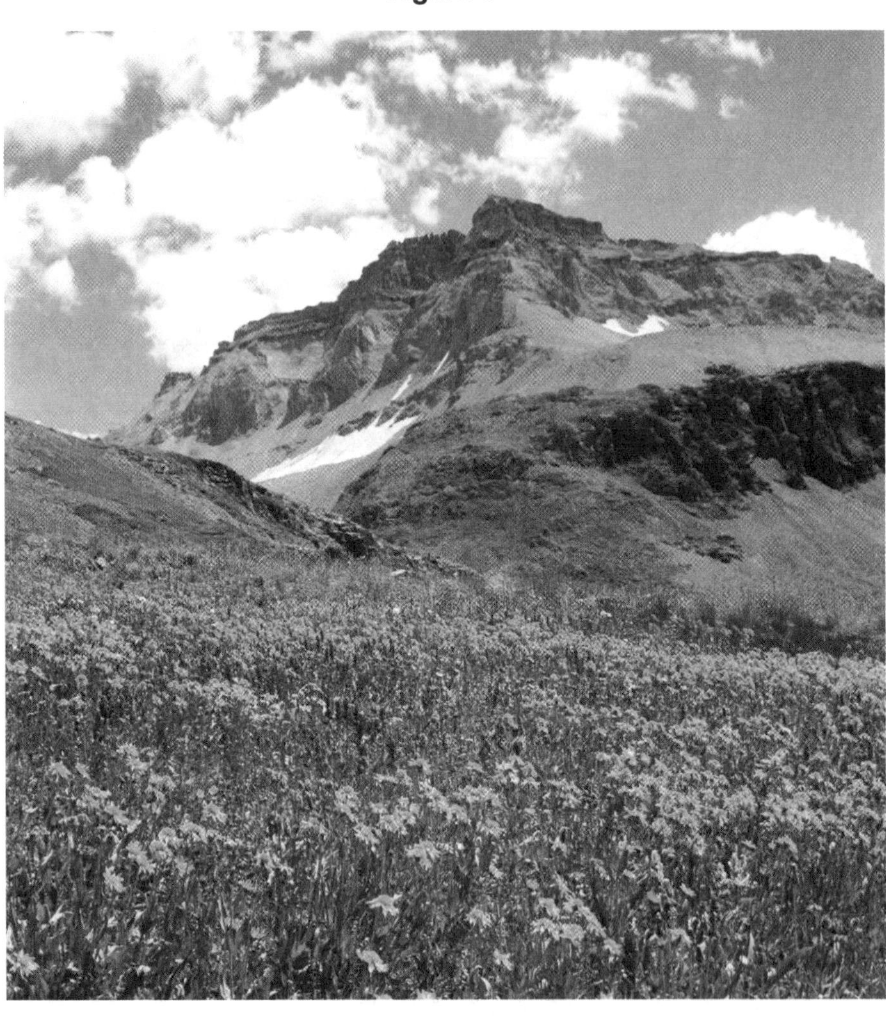

a) Give **three** facts about the vegetation of cold environments shown in **Figure 6**.

[3 marks]

___

___

___

___

___

___

Study **Figure 7** showing three climate graphs

Figure 7

A

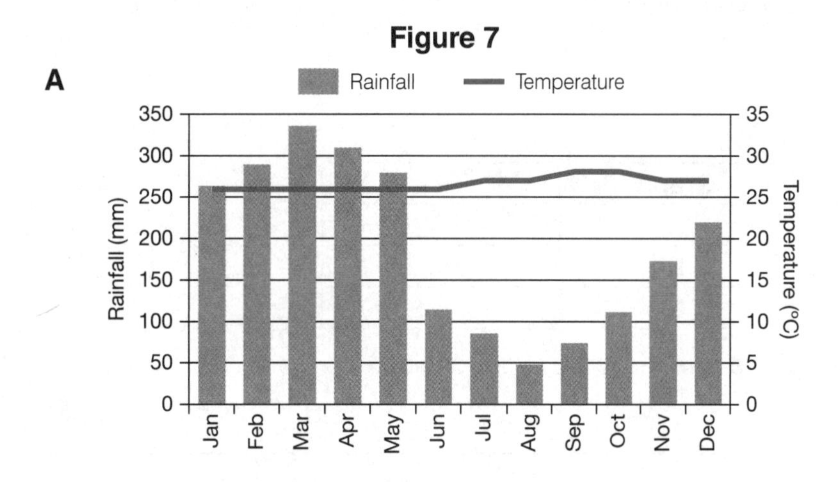

B

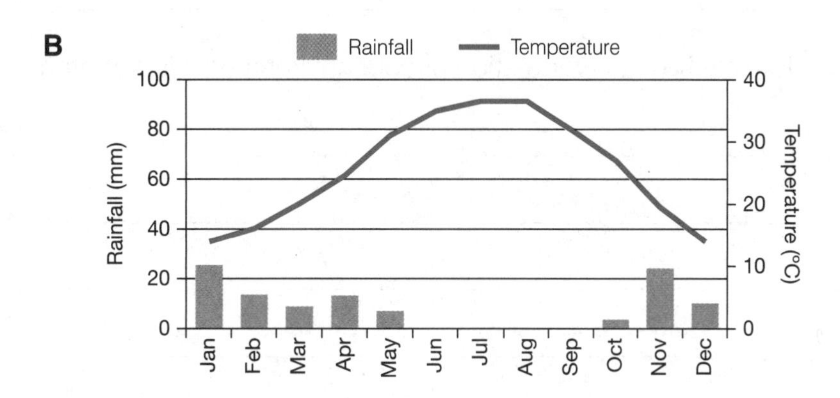

C

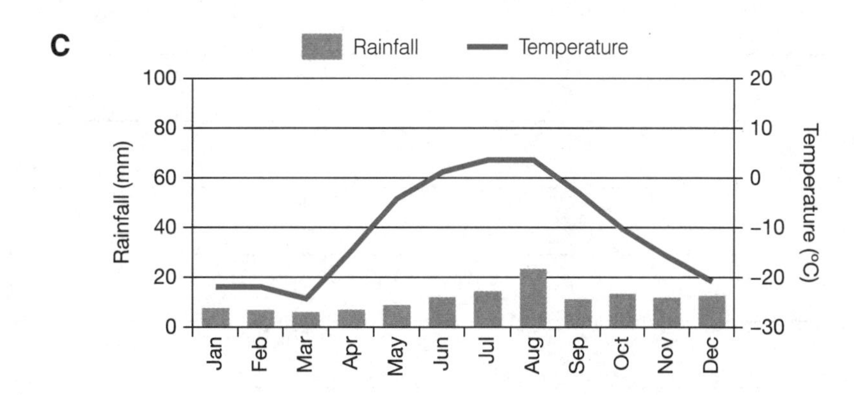

b) Which of the graphs **A**, **B** or **C** in **Figure 7** represents the climate of a cold environment?

Tick the box showing the correct answer.

[1 mark]

Graph	Tick
A	
B	
C	

c) State **one** reason for your choice.

[1 mark]

d) Explain how the vegetation of cold environments is adapted to the climate.

[6 marks]

Study **Figure 8**, showing the amount of deforestation that has taken place in the tropical rainforest of Amazonia in recent years.

Figure 8

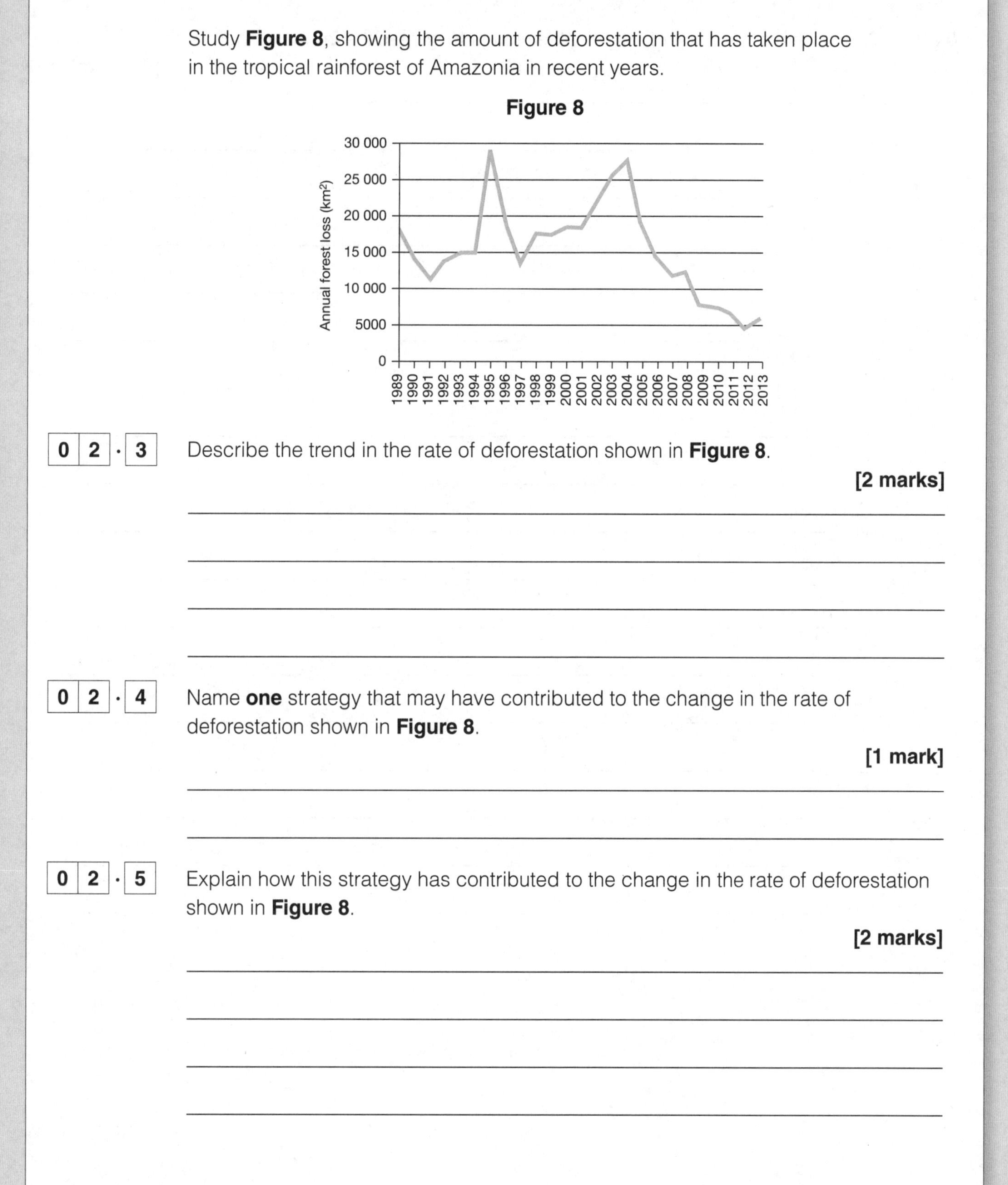

0 2 · 3 Describe the trend in the rate of deforestation shown in **Figure 8**.

[2 marks]

0 2 · 4 Name **one** strategy that may have contributed to the change in the rate of deforestation shown in **Figure 8**.

[1 mark]

0 2 · 5 Explain how this strategy has contributed to the change in the rate of deforestation shown in **Figure 8**.

[2 marks]

0 2 · 6 Use a case study of a tropical rainforest to assess the impact of deforestation.

[9 marks]

Section C Physical landscapes in the UK

Answer **two** questions from the following:
Question 3 (Coasts), Question 4 (Rivers), Question 5 (Glacial)

Question 3 Coastal landscapes in the UK

Study **Figure 9**, photographs that show different types of sea defence.

Figure 9

A

B

C

D

0 3 · 1 **Figures 9 A–D** are examples of which engineering strategy?
Tick the correct answer.

[1 mark]

Engineering strategy	Tick
Hard engineering	
Soft engineering	

0 3 · 2 Write the correct letter **(A–D)** against each example of a sea defence.

[3 marks]

Sea defence type	Letter
Gabions	
Groynes	
Rock armour	
Sea wall	

0 3 · 3 New methods of sea defence involve managed retreat.
Explain why this management strategy is considered to be more acceptable than the methods illustrated in **Figure 9**.

[4 marks]

0 3 · 4 Which one of the following coastal landforms forms largely as a result of erosion?
Circle the correct letter.

[1 mark]

A Beach **B** Bay **C** Spit **D** Bar

0 3 · 5 With the aid of a diagram explain the formation of cliffs and wave-cut platforms.
(Sketch your diagram on a separate piece of paper.)

[6 marks]

Question 4 **River landscapes in the UK**

Study **Figure 10**, photographs that show different forms of river management strategies.

Figure 10

A **B**

C  **D**

0 4 · 1 **Figures 10 A–D** are examples of which management strategy? Tick the correct answer.

[1 mark]

Management strategy	Tick
Hard engineering	
Soft engineering	

0 4 · 2 Write the correct letter **(A–D)** against each example of a flood management strategy.

[3 marks]

Flood management strategy	Letter
Dams and reservoirs	
Embankments	
Flood relief channel	
River straightening	

0 4 · 3 New methods of flood control involve floodplain zoning.
Explain why this management strategy is considered to be more acceptable than the methods illustrated in **Figure 10**.

[4 marks]

0 4 · 4 Which one of the following river landforms forms largely as a result of erosion?
Circle the correct letter.

[1 mark]

A Flood plain **B** Levee **C** Interlocking spurs **D** Estuary

0 4 · 5 With the aid of a diagram explain the formation of an ox-bow lake.
(Sketch your diagram on a separate piece of paper.)

[6 marks]

Question 5 **Glacial landscapes in the UK**

Study **Figure 11**, photographs that show different strategies used to manage the impact of tourism in a glaciated upland area.

Figure 11

A

B

C

D

0 5 · 1 Which of the following impacts of tourism do **Figures 11 A–D** try to overcome? Tick the correct answer.

[1 mark]

Type of impact	Tick
Economic	
Environmental	

0 5 · 2 Write the correct letter **(A–D)** against each example of a tourism impact strategy.

[3 marks]

Tourism impact strategy	Letter
Footpath erosion repair	
Out of season tourism	
Use of public transport	
Walking trails	

0 5 · 3 Explain why the development of tourism in a glaciated upland area can cause conflict.

[4 marks]

0 5 · 4 Which one of the following glacial landforms forms largely as a result of erosion? Circle the correct letter.

[1 mark]

A Hanging valley **B** Drumlin **C** Erratic **D** Lateral moraine

0 5 · 5 With the aid of a diagram explain the formation of a corrie.
(Sketch your diagram on a separate piece of paper.)

[6 marks]

GCSE 9-1 Geography AQA
Practice Paper

Paper 2 Challenges in the human environment

Time allowed: 1 hour 30 minutes
Total number of marks: 88 (including 3 marks for spelling, punctuation, grammar and specialist terminology (SPaG))

Instructions

Answer **all** questions in Section A and Section B
Answer question 3 and **one** from questions 4, 5 or 6

Section A Urban issues and challenges

Answer **all** questions in this section.

Question 1 Urban issues and challenges

Study **Figure 1**, a map showing the growth of cities in Africa.

Figure 1

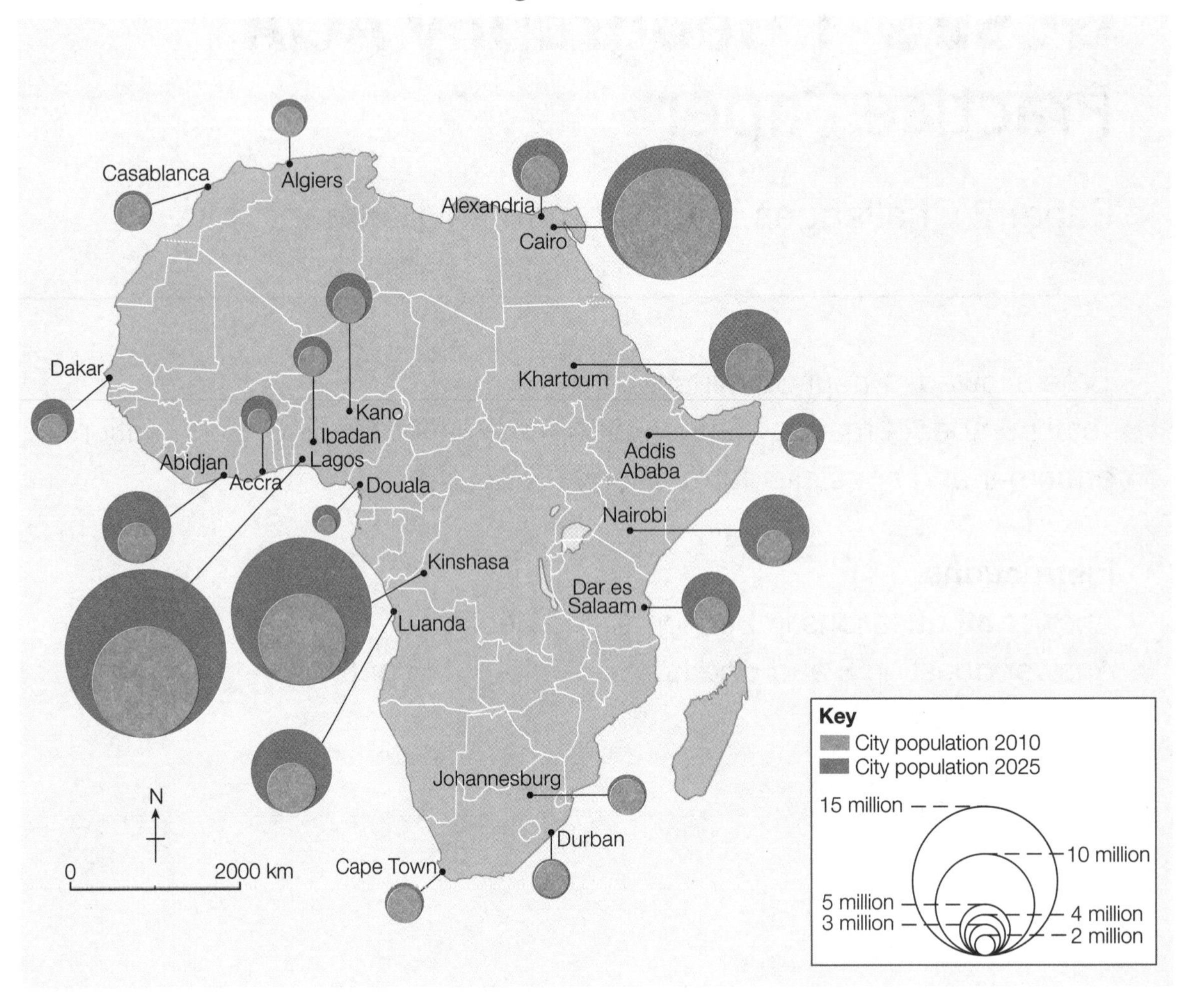

0 1 · 1 Name the African city which will have the largest population in 2025.

[1 mark]

____

____

0 1 · 2 By how much will the population of Cairo have increased between 2010 and 2025?

[1 mark]

____

____

0 1 · 3 Which African city will have seen the greatest percentage increase between 2010 and 2025?

[1 mark]

____

____

0 1 · 4 How many African cities will be megacities in 2015? Circle the correct answer.

[1 mark]

1 2 3 4

0 1 · 5 African countries are **either** LICs **or** NEEs. Explain why a growing percentage of Africans live in urban areas.

[4 marks]

____

____

____

____

____

____

____

____

Study **Figure 2**, the Green City Index and also **Figure 3**, a comparison of London and Copenhagen using the Green City Index. The Green City Index is used to measure the sustainability of cities. A high index indicates greater sustainability.

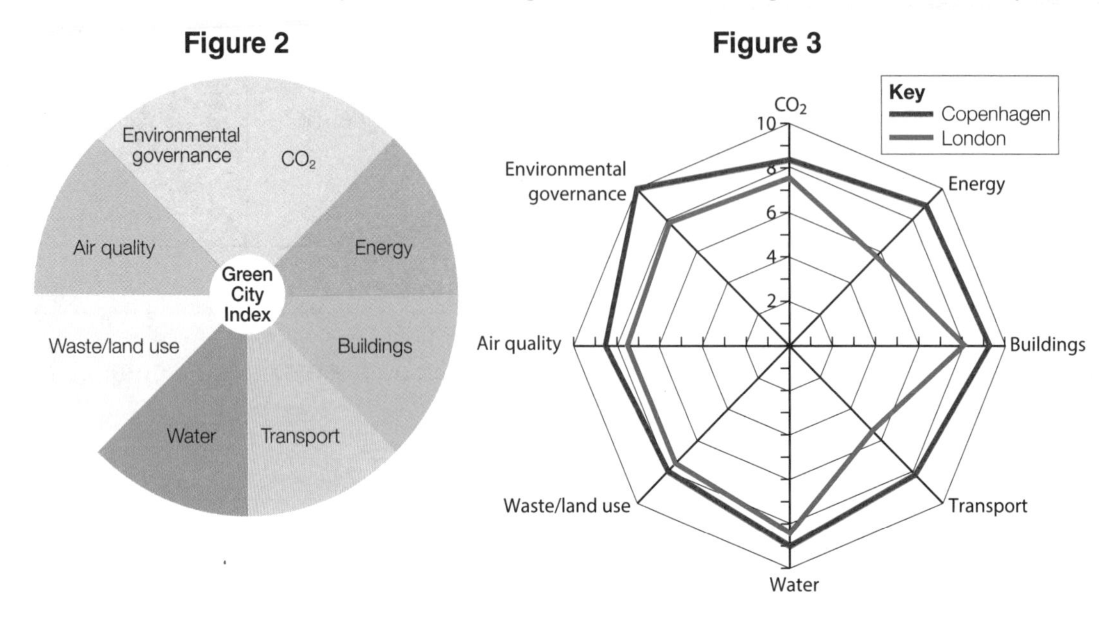

0 1 · 6 Using **Figures 2** and **3** and your own knowledge, justify why Copenhagen is considered to be a more sustainable city than London.

[6 marks]

0 1 · 7 What is meant by the term 'urban deprivation'?

[1 mark]

0 1 · 8 Using your case study of a major city in the UK, assess the challenges faced because of social and economic inequalities in different parts of the city.

[9 marks]
[+3 SPaG marks]

Name a major city in an LIC or NEE that you have studied.

Name of city ______________________

0 1 · 9 Name **two** environmental issues caused by the urban growth of your chosen city.

[2 marks]

Issue 1 ____

____

Issue 2 ____

____

0 1 · 10 Explain how your chosen city has attempted to improve its environment.

[4 marks]

____

____

____

____

____

____

____

____

Section B The changing economic world

Answer **all** questions in this section

Question 2 The changing economic world

| 0 | 2 | · | 1 | What do the letters HDI stand for? **[1 mark]**

| 0 | 2 | · | 2 | Explain why the HDI is a good measure of a country's level of development. **[2 marks]**

Study **Figure 4**, a scattergraph showing the correlation between HDI and birth rate.

Figure 4

| 0 | 2 | · | 3 | Draw in the best-fit line on **Figure 4**. **[1 mark]**

| 0 | 2 | · | 4 | What is the correlation between GNI per capita and birth rate shown on **Figure 4**? **[1 mark]**

Study **Figure 5**, two photographs that show changes that have occurred in the South Yorkshire town of Rotherham.

Figure 5

Former Dinnington Colliery, South Yorkshire | Dinnington Business Centre on site of former Dinnington Colliery

0 2 · 5 In what ways are the changes shown in **Figure 5** typical of economic change in the UK?

[4 marks]

0 2 · 6 Explain how globalisation and government policies have caused economic change in the UK.

[6 marks]

0 2 · 7 Name **two** recent transport improvements that have been undertaken in the UK in recent years.

[2 marks]

Improvement 1 ______

Improvement 2 ______

0 2 · 8 Explain how improvements and new developments in transport affect regional growth in the UK.

[4 marks]

0 2 · 9 Using your study of an LIC or NEE, evaluate the effects of economic development on the quality of life for the country's population.

[9 marks]

Section C The challenge of resource management

Answer Question 3 (Resources) and **either** Question 4 (Food) **or** Question 5 (Water) **or** Question 6 (Energy)

Question 3 Resource management

Study **Figure 6**, a pie chart showing the use of water in the UK in 1997.

Figure 6

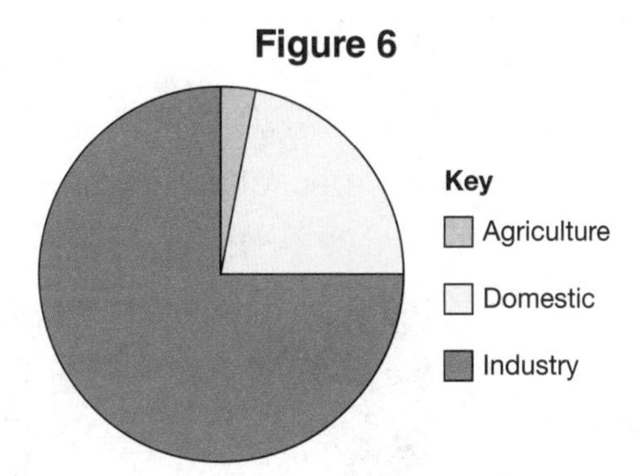

0 3 . 1 What percentage was used by industry?

[1 mark]

____

0 3 . 2 If a pie chart was drawn to show the use of water in the UK today, which of the following is it most likely to show? Tick the most appropriate answer.

[1 mark]

Change from Figure 6	Tick
Agricultural use no change, domestic use decreased, industrial use increased	
Agricultural use gone down, domestic use increased, industrial use increased	
Agricultural use increased, domestic use increased, industrial use decreased	

0 3 . 3 Give **two** reasons for your choice.

[2 marks]

____

____

____

____

0 3 · 4 Describe how pollution management helps to maintain sufficient supplies of water in the UK.

[4 marks]

Study **Figure 7**, which shows a packet of Kenyan vegetables.

Figure 7

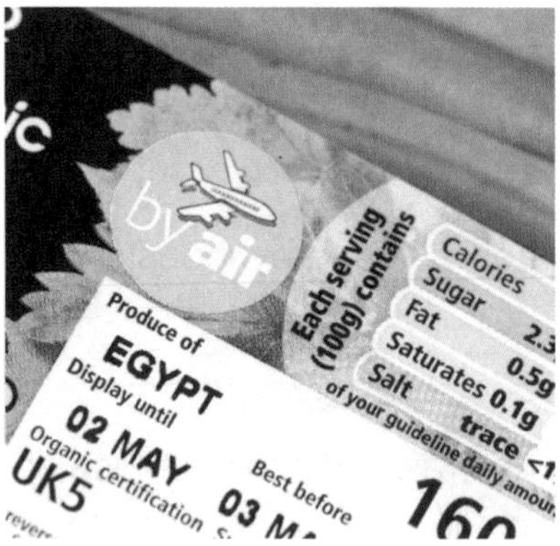

0 3 · 5 With the aid of **Figure 7** and your own knowledge, explain why changes in people's eating habits have meant that an increasing amount of food has to be imported into the UK.

[6 marks]

Answer **either** Question 4 (Food) **or** Question 5 (Water) **or** Question 6 (Energy)

Question 4 **Food**

Study **Figure 8**, a diagram where the size of a country is drawn in proportion to the number of people that live there who suffer from undernutrition.

Figure 8

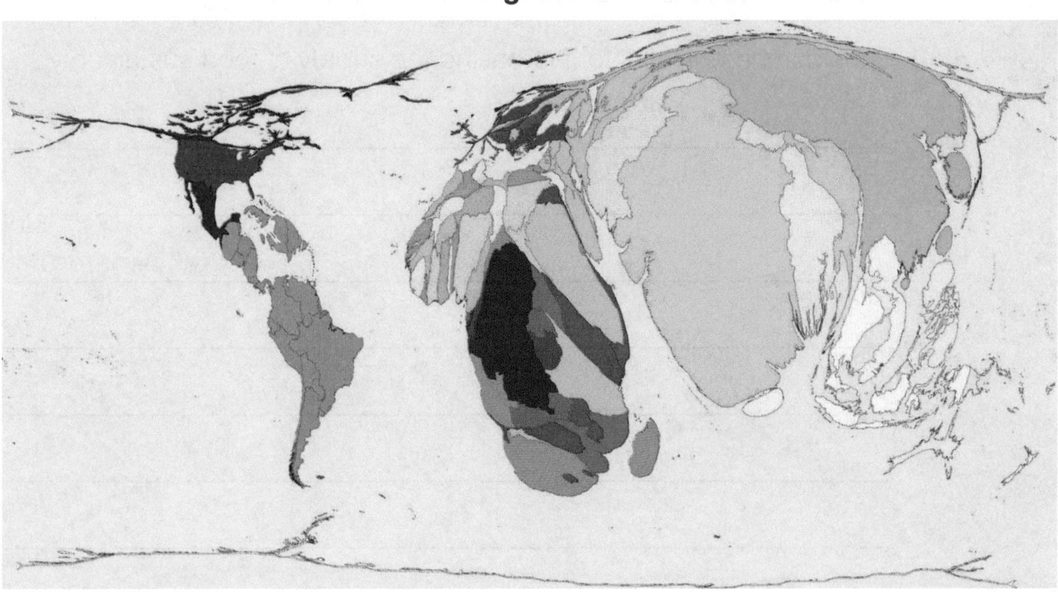

0 4 . 1 Which continent appears to have the highest proportion of its population suffering from undernutrition?

[1 mark]

____

0 4 . 2 Describe the pattern of undernutrition shown in **Figure 8**.

[2 marks]

____

____

____

____

0 4 · 3 Give **two** reasons for the increasing global consumption of food.

[2 marks]

Reason 1 ____

____

Reason 2 ____

____

0 4 · 4 Consider the potential for increasing the supply of food sustainably.

[6 marks]

____

____

____

____

____

____

____

____

____

____

____

____

Question 5 **Water**

Study **Figure 9**, a diagram which shows use of a country's own internal water resources. Each country is drawn in proportion to how much of its own renewable water resources it uses.

Figure 9

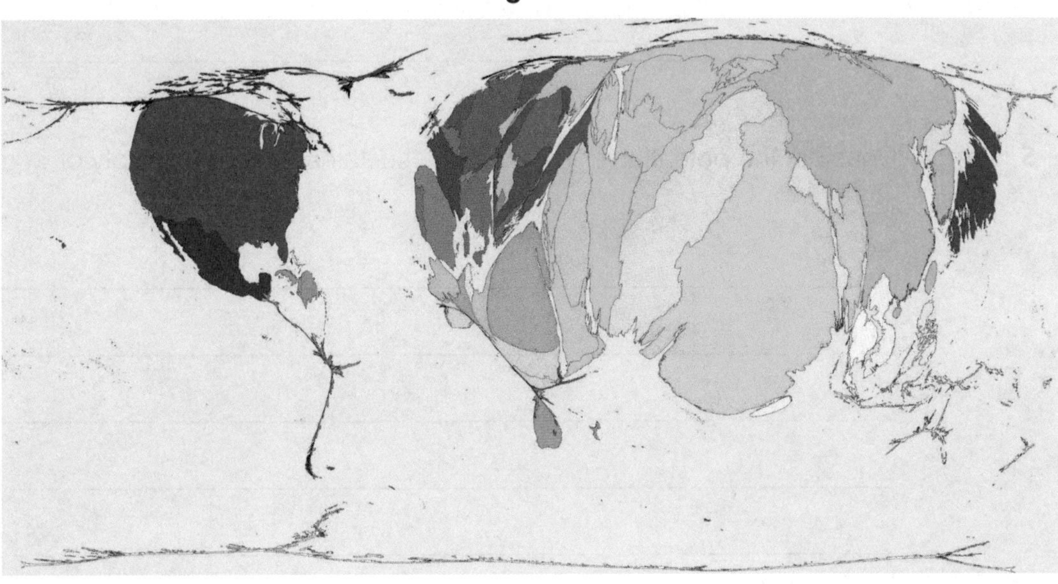

0 5 · 1 Which continent appears to use the greatest amount of its internally available water?

[1 mark]

____

0 5 · 2 Describe the pattern of water usage shown in **Figure 9**.

[2 marks]

____

____

____

____

0 5 · 3 Give **two** reasons for the increasing global consumption of water.

[2 marks]

Reason 1 ____________________

____

Reason 2 ____________________

____

0 5 · 4 Consider the potential for producing a sustainable future supply of water.

[6 marks]

____

____

____

____

____

____

____

____

____

____

____

____

Question 6 **Energy**

Study **Figure 10**, a diagram where the size of a country is drawn in proportion to the rate at which its own reserves of fossil fuels are being used up.

Figure 10

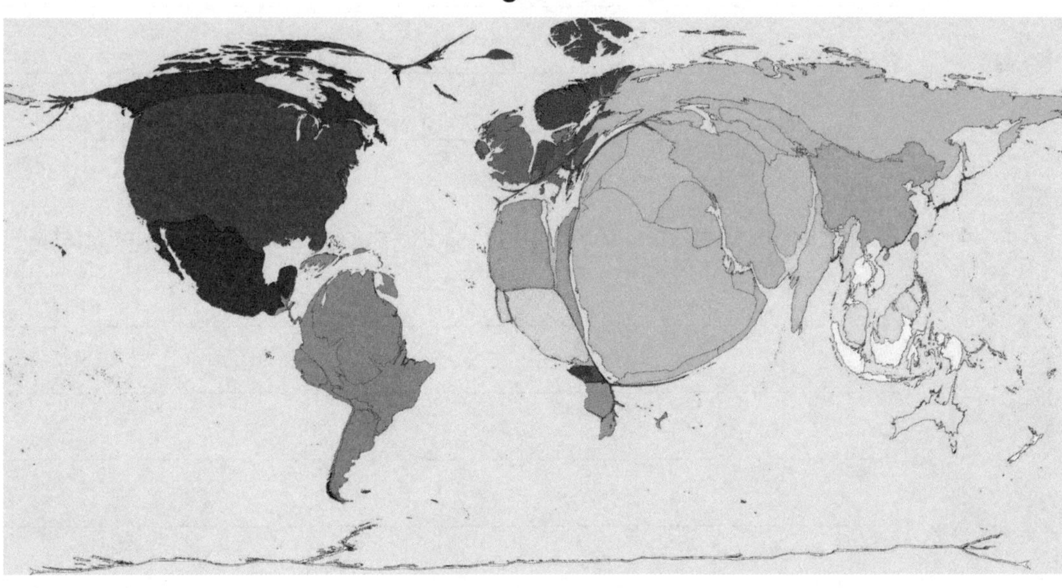

0 6 · 1 Which continent appears to be using its reserves of fossil fuels at the greatest rate?

[1 mark]

___

0 6 · 2 Describe the pattern shown in **Figure 10**.

[2 marks]

___

___

___

___

0 6 · 3 Give **two** reasons for the increasing global energy consumption.

[2 marks]

Reason 1 ____

____

Reason 2 ____

____

0 6 · 4 Consider the potential for increasing the supply of energy sustainably.

[6 marks]

____

____

____

____

____

____

____

____

____

____

____

____

GCSE 9-1 Geography AQA
Practice Paper

Paper 3 Geographical applications

Time allowed: 1 hour 15 minutes
Total number of marks: 76 (including 6 marks for spelling, punctuation, grammar and specialist terminology (SPaG))

Instructions
Answer **all** questions
Use a clean copy of the pre-release resources booklet

Section A Issue evaluation

Answer **all** questions in this section

Study **Figure 1** in the resources booklet, 'People living in a high-risk earthquake zone'.

0 1 · 1 Name the **two** cities which, if they suffer a severe earthquake, would have the greatest international impact.

[1 mark]

City 1 ______________________ **City 2** ______________________

0 1 · 2 Suggest why this is the case.

[2 marks]

____

____

____

0 1 · 3 Suggest why people continue to live in areas with a high earthquake risk.

[4 marks]

____

____

____

____

____

____

0 1 · 4 The scattergraph suggests there is no correlation between the strength of an earthquake and the value of the property damage. These earthquakes occurred in both LICs and HICs. Suggest how this can be an explanation of the pattern shown on the scattergraph.

[2 marks]

____

____

____

____

Study **Figure 2** in the resources booklet, 'Turkey's earthquake risk'.

0 2 · 1 Describe the distribution of earthquakes in Turkey.

[2 marks]

0 2 · 2 Look at the map and explain the sequence of the earthquakes in Turkey.

[2 marks]

0 2 · 3 'The situation appeared chaotic as night fell.'

Use **Figure 2** and your own understanding to discuss the validity of this statement about the Van earthquake.

[6 marks]

Study **Figure 3** in the resources booklet, 'Istanbul'.

0 3 · 1 A recent newspaper article had the following heading.

A disaster waiting to happen

Justify that Istanbul is likely to suffer an earthquake disaster in the near future.

[6 marks]

Three options have been suggested to plan for reducing the impact of an earthquake hitting Istanbul.

Option 1	Have strict building codes which insist that all new buildings must be constructed to strict earthquake-proof standards and make older buildings as earthquake proof as possible.
Option 2	Develop a city plan where all emergency services, such as hospitals, fire stations and power stations, are built as far away from the fault line as possible.
Option 3	Develop a new range of monitoring stations and ensure that the population is made fully aware of earthquake drills and evacuation procedures.

0 3 · 2 Choose the option which you think will be most effective in reducing deaths and injuries in the event of a major earthquake hitting Istanbul.

Use evidence from the Resource booklet and your own understanding to explain why you have reached this decision.

[9 marks] [+ 3 SPaG marks]

Section B Fieldwork

Answer **all** questions

Study **Figure 4**, a graph showing some results from a geographical investigation in an urban area.

Figure 4

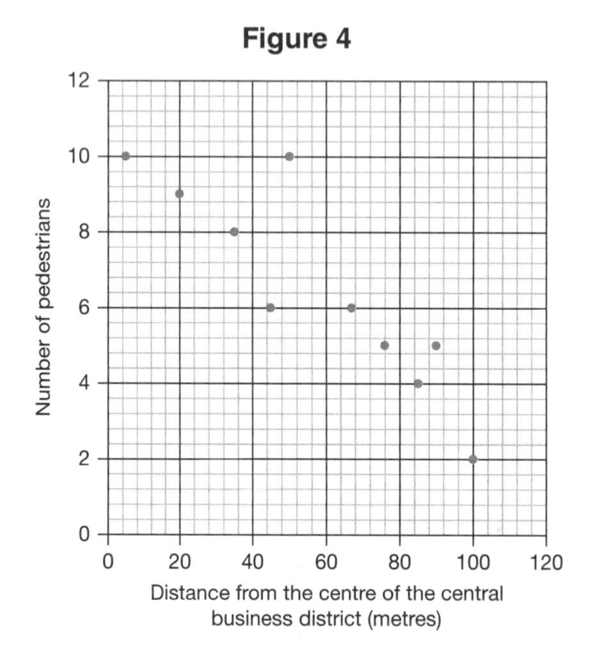

0 4 . 1 Three pedestrians were counted 115 metres from the centre of the central business district. Add this information to the graph in **Figure 4**.

[1 mark]

0 4 . 2 Draw in the best-fit line on **Figure 4**.

[1 mark]

0 4 . 3 One of the following statements is correct. Tick the correct one.

[1 mark]

Statement	**Tick if correct**
The dependant variable is on the *x*-axis	
The dependant variable is on the *y*-axis	

Study **Figure 5**, a graph for one area of Birmingham, that shows the population by ethnicity taken from the 2001 census and also from a 2017 sample.

Figure 5

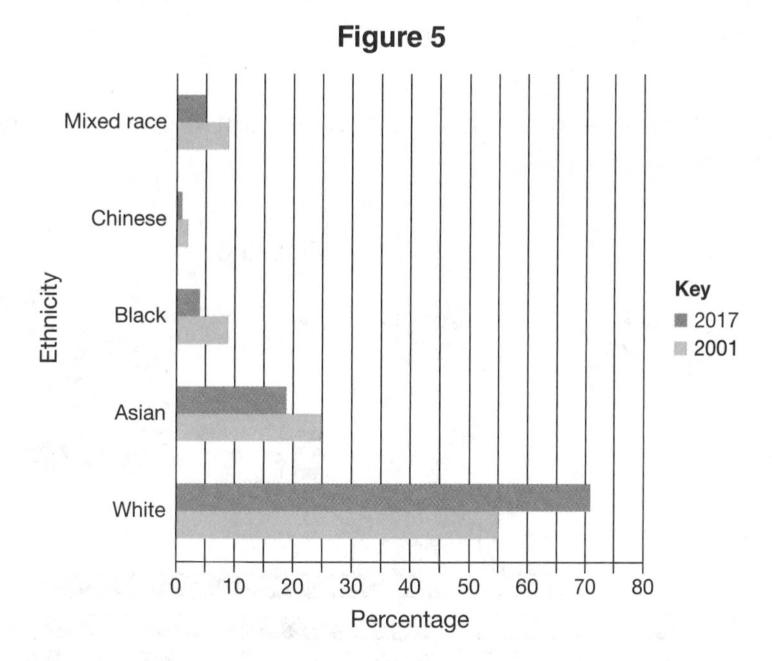

0 4 · 4 What percentage of the population was white in the 2001 census?

[1 mark]

____

0 4 · 5 How has the population changed between 2001 and 2017?

[2 marks]

____

____

____

0 4 · 6 Explain why it is necessary to use both primary and secondary data to show more accurately how the population is changing.

[4 marks]

____

____

____

____

____

____

____

____

Study *either* **Figure 6** *or* **Figure 7**.

Figure 6 shows equipment that could be used for a physical geography investigation at the coast.

Figure 7 shows equipment that could be used for a physical geography investigation on a river.

Figure 6

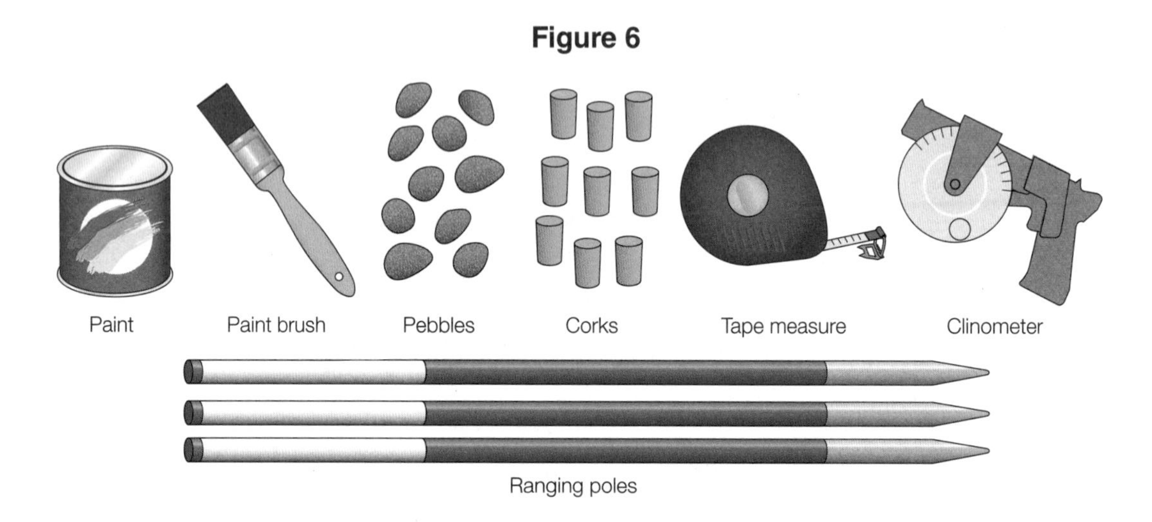

Figure 7

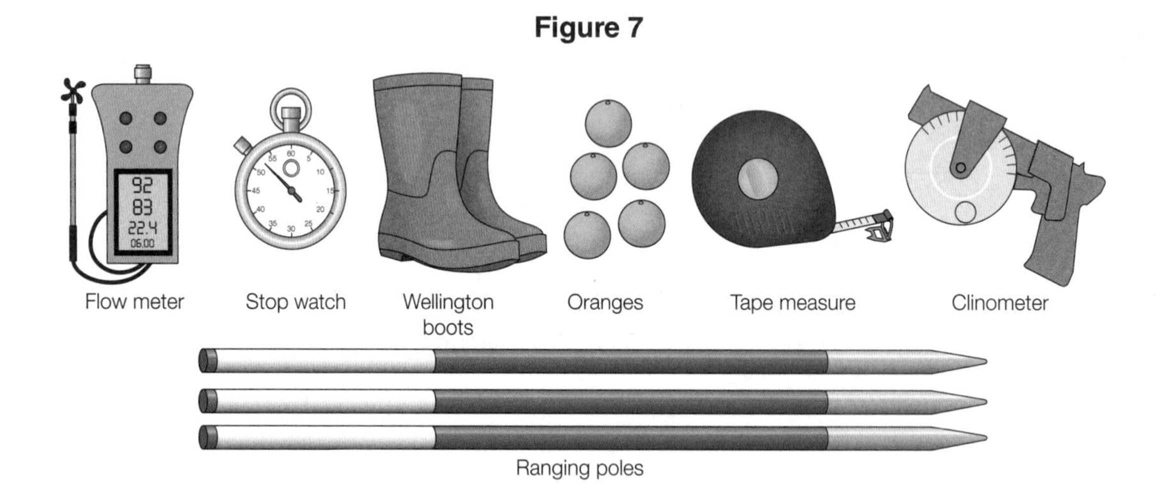

0 4 . 7 Circle the physical geography investigation you have chosen.

Coast River

Choose **two** of the pieces of equipment shown in *either* **Figure 6** *or* **Figure 7** and explain why they are useful to collect data, from which conclusions can be drawn.

[4 marks]

Piece of equipment 1

____

____

____

____

Piece of equipment 2

____

____

____

____

Study **Figure 8**, a graph showing the change in velocity over the course of a river. The results form part of a river study.

Figure 8

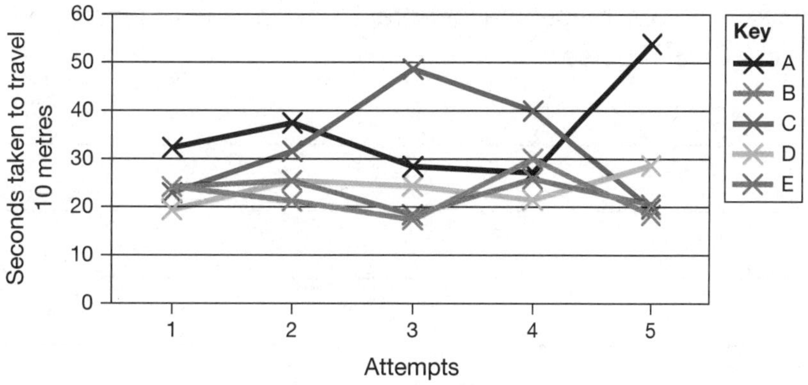

0 4 . 8 State why this is an inappropriate method of data presentation for this type of data.

[1 mark]

____

____

0 4 . 9 Name a more appropriate form of presentation to illustrate this data.

[1 mark]

____

____

0 5 · 1 State the title of your fieldwork enquiry in which **human** geography data was collected.

Title of fieldwork enquiry ____________________

Explain the advantages of the location(s) used for your fieldwork enquiry.

[2 marks]

0 5 · 2 Explain how this fieldwork enquiry helped your understanding of a geographical concept or theory studied as part of your GCSE course.

[3 marks]

0 5 · 3 How far did the conclusions you reached in this fieldwork enquiry support or disagree with the general geographical concept or theory you studied as part of your GCSE course?

[6 marks]

0 5 · 4 State the title of your fieldwork enquiry in which you investigated the interaction between physical and human geography.

Title of fieldwork enquiry ______________________________

To what extent did your fieldwork enquiry show the links between physical and human geography?

[9 marks]

Figure 1

The tropical rainforest global ecosystem or biome

The climate of the tropical rainforests

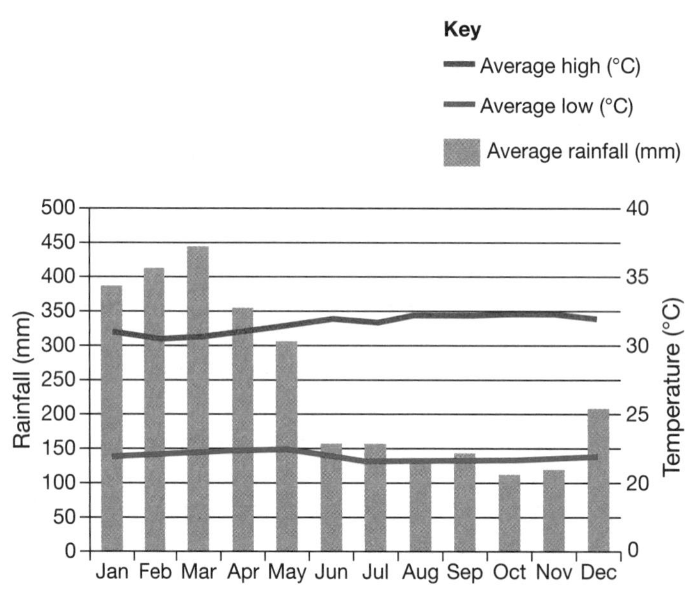

Climate graph for Belem, Brazil

The sun is high throughout the year without any definite seasons. Belem's annual average temperature is 32°C, with over 2900 mm of rainfall and 2200 hours of sunshine each year – an average of 6 hours a day. Humidity remains high throughout the year.

The effect of climate change

A major threat to tropical rainforests is global warming. Rising populations and increasing resource consumption add greenhouse gases to the atmosphere, which causes the climate to change. Some scientists think global warming will lead to species extinction at an unprecedented rate.

Temperature rise	Impact on species	Impact on biome
3°C	20–50% of species face extinction	• Forest gets stressed by drought • Increased danger of fire • Flooding causes the loss of mangrove • Pests and diseases thrive in rising temperatures

Impact of global warming on tropical rainforests

The rainforest suffers from climate change.

- During drought, the forests stop absorbing carbon dioxide and emit it instead, because plants stop growing and therefore can no longer absorb the gas.
- Forest fires break out in the drought conditions, burning trees and releasing carbon dioxide.
- Leaf litter dries up, so decomposer organisms die out, threatening the nutrient cycle.
- Leaves in the canopy die, reducing the availability of food which affects food webs.
- Deforestation makes droughts more common and more severe.
- With fewer trees, there is less evaporation and transpiration. This means there are fewer clouds and less rain.

The threat caused by deforestation

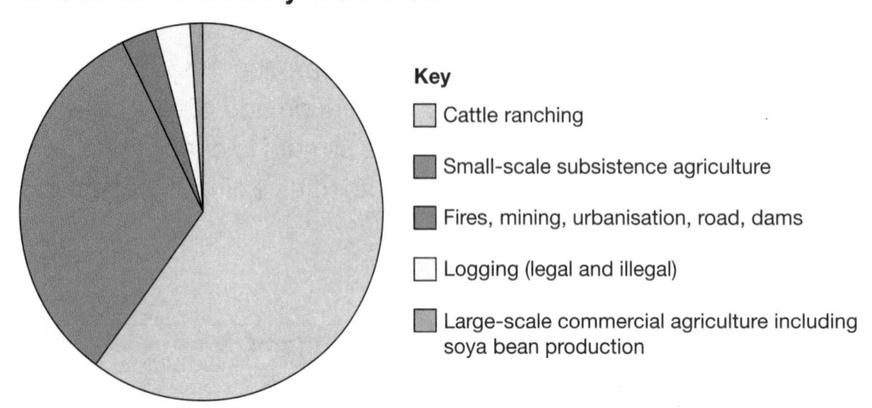

Causes of deforestation in Amazonia

Causes of deforestation include:

- **poverty:** in many LICs (low-income countries), local people cut down small areas of forest for land to farm because they have no other way of making a living.
- **debt:** countries are driven to cut down forests, export timber or grow cash crops to pay off debts.
- **economic development:** most tropical forests are in the developing world. In order to develop their economies, forest is sacrificed to be replaced by roads, expanding cities, and rivers dams for hydroelectric power (HEP) stations.
- **demand for resources:** tropical forests contain raw materials. These include timber and also oil, gas, iron ore and gold. To reach these resources, forest has to be destroyed. Land is also needed to feed growing populations.

Country	Annual rate of deforestation (%)
Burundi	−4.7
Nigeria	−3.1
Indonesia	−1.9
Malaysia	−0.7
Brazil	−0.6
Democratic Republic of Congo	−0.2

Rates of deforestation in six countries

Figure 2

Biofuels in Indonesia

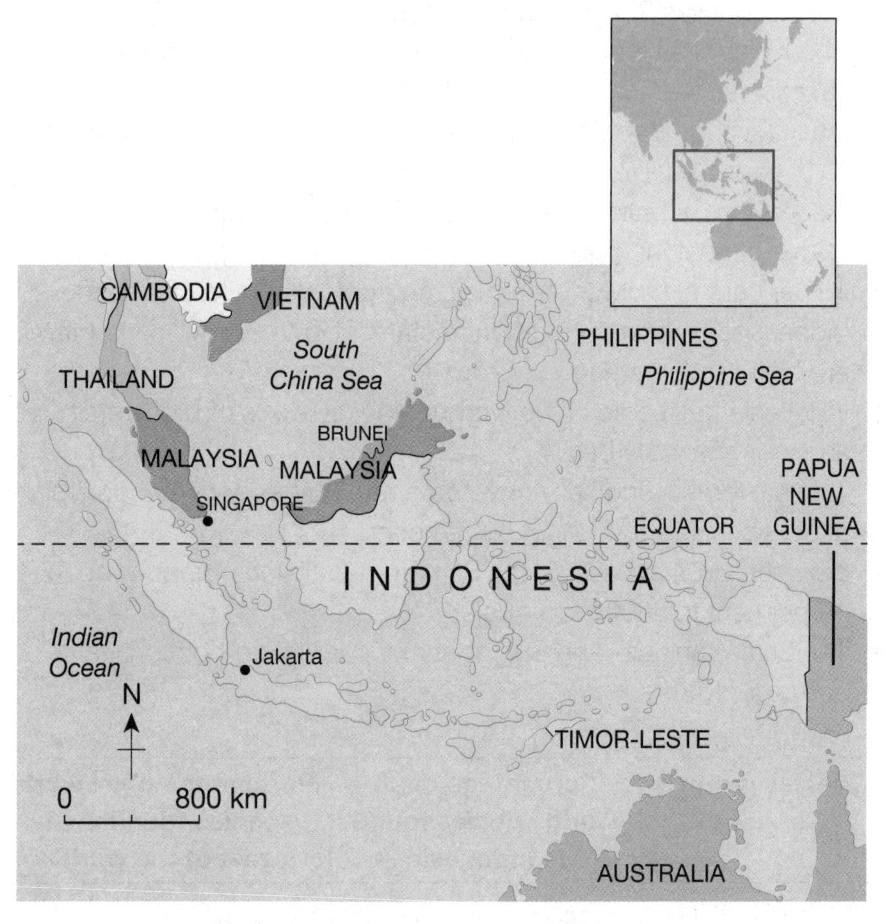

The location of Indonesia in south-east Asia

The issue: biofuels in Indonesia's rainforests

- Indonesia's population and economy have each grown rapidly in recent years.
- Some of Indonesia's economic growth has been based on deforesting tropical rainforests. The cleared land is used to grow oil palm plants.
- Indonesia's government sees this as a way to develop the country. They hope it will lift rural people out of poverty and allow major companies to profit from palm oil.
- Indonesia's tropical rainforests have very high levels of biodiversity, but are remote and hard to monitor.

Getting to know Indonesia

- Indonesia is a large country in Asia (see map on page 159). It is spread over numerous large and small islands whose total population was 252 million people in 2015.
- About half of all Indonesians live in urban areas. The capital city, Jakarta, is located on the island of Java.
- It has a young population. 27 per cent of the population is aged 0–14 years, but life expectancy is quite high at 71 years.
- Incomes in Indonesia are US$3800 per person per year and 39 per cent of the population still works in farming.
- 10 per cent of the world's remaining tropical rainforests are found in Indonesia, covering 98 million hectares. Up to 2.5 million hectares are being deforested every year.
- Indonesia has some of the world's highest levels of biodiversity, as shown in the table below.
- Tropical forests originally covered about 85 per cent of Indonesia although this is now about 50 per cent.
- Between 2004 and 2008, the Sumatran orangutan population fell by 14 per cent to 6600 individuals.
- The Sumatran tiger and Sumatran rhinoceros are both critically endangered.

Species group	Percentage of all world species found in Indonesia	Percentage of all world species found in the rest of the world
Plant species	10	90
Mammal species	12	88
Reptiles & amphibians	16	84
Birds	17	83
Fish	25	75

Biodiversity in Indonesia compared to the rest of the world

Tropical rainforest in Indonesia

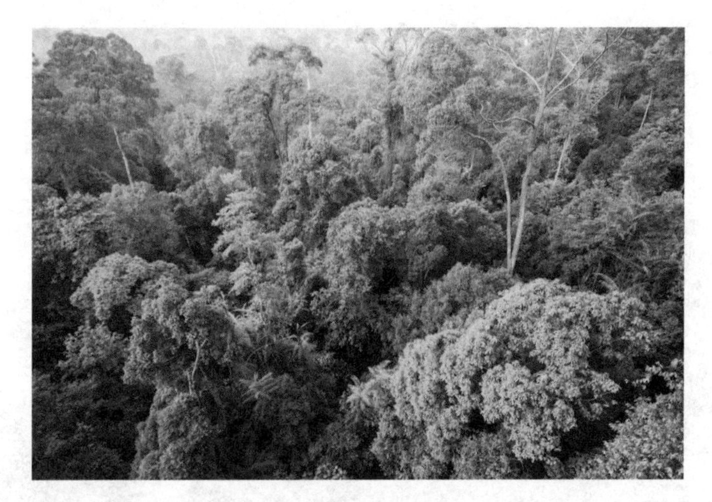

- Much of the tropical rainforest in Indonesia is found in mountainous areas.
- Different species are found at different altitudes.
- The endangered orangutan is found on Sumatra and Borneo.
- Deforestation threatens even isolated areas of tropical forest.

Peatland swamp forest in Indonesia

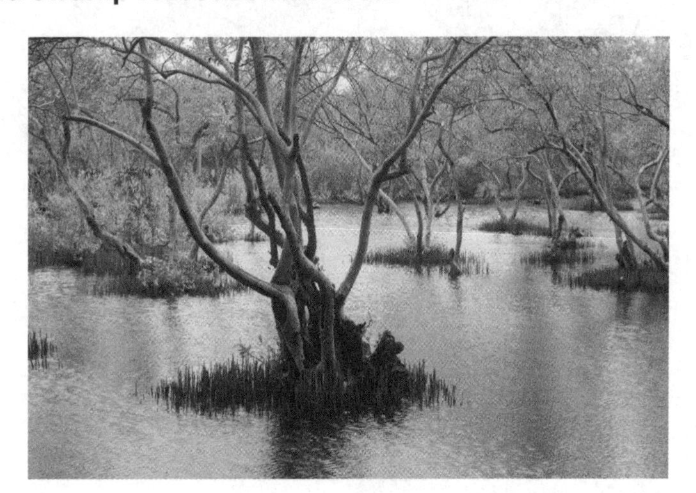

- In lowland areas the forest grows in waterlogged swamps.
- Beneath the forest is a layer of peat (waterlogged soil).
- Peat is a huge carbon store.
- When the peatland forest is cleared, the peat begins to decompose, releasing its stored carbon.

Figure 3

The threat to the Indonesian rainforests

The Indonesian palm oil industry

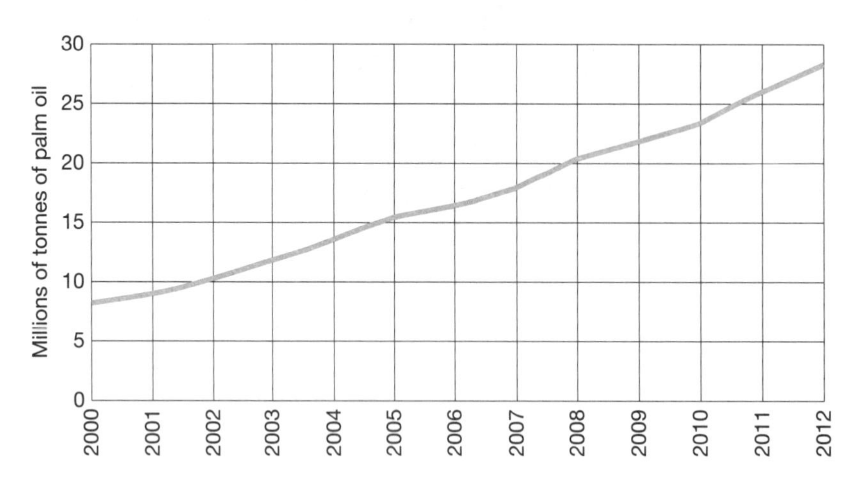

Palm oil production in Indonesia 2000–12

- Indonesia's tropical rainforests are being cleared to create oil palm plantations.
- In 2011, oil palm plantations covered 7.8 million hectares in Indonesia.
- Between 2008 and 2020, Indonesia plans to double production of palm oil to 40 million tonnes per year and increase oil palm plantations by 4 million hectares.
- High demand for palm oil from emerging economies in Asia is driving this increase.
- About 75 per cent of oil palm plantations estates are located on Sumatra and Kalimantan.
- About 50 per cent of plantations are small, family-run farms.
- Over 2 million people are employed in Indonesia's palm oil industry.
- In 2010, palm oil made up 7 per cent of Indonesia's exports, valued at US$12 billion.
- Palm oil is used to make biodiesel (a replacement for diesel from crude oil), food products, shampoo and lipstick. It can only be grown in tropical areas like Indonesia.

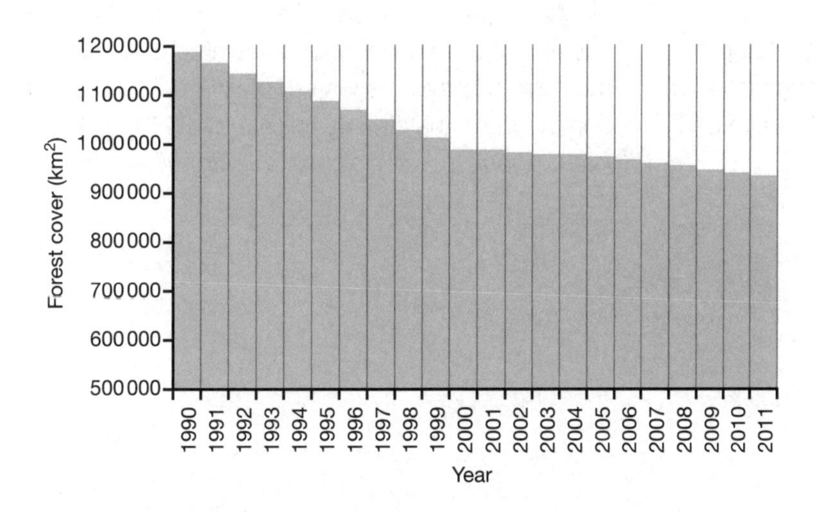

The decline in Indonesian rainforests 1990–2011 (km²)

Contrasting views about palm oil development in Indonesia

Organisation	View
WWF is an environmental pressure group and NGO.	'Large areas of tropical forests have been cleared to make room for vast oil palm plantations – destroying habitats for many endangered species, including rhinos, elephants and tigers. In some cases, the expansion of plantations has led to the eviction of forest-dwelling people.' – *From the WWF article 'Environmental and Social Impacts of Palm Oil Production' at wwf.panda.org*
World Growth is a pressure group that promotes globalisation.	'Palm oil provides developing nations and the poor with a path out of poverty. Expanding sustainable agriculture such as oil palm plantations provides plantation owners and their workers with a means to improve their standard of living.' – *From the World Growth report 'The Economic Benefit of Palm Oil to Indonesia', Feb 2011*
Cargill is a TNC based in the USA that grows, processes and sells palm oil.	'Millions of people around the world depend on palm oil. We believe that palm oil should be produced sustainably. We have made a commitment that the palm oil products we supply will be certified as coming from sustainable forests by 2020.' – *Adapted from various Cargill policies on palm oil at www.cargill.com*

Paper 3 Geographical applications

Figure 1

People living in a high-risk earthquake zone

World Cities at risk from a serious earthquake

City (country)	Population (millions)	Estimate of deaths if earthquake of 6+ on Richter scale occurs	Other Comments
Tokyo (Japan)	29.4	9000	Situated on 'Pacific Ring of Fire'. Would have a global effect because of the city's importance as an international financial centre.
Djakarta (Indonesia)	17.7	11000	Situated on the 'Pacific Ring of Fire', with 50% of the city below sea level. Built on soil which could liquefy during an earthquake.
Manila (Philippines)	16.8	13000	One of the most densely populated cities in the world. Situated on the 'Pacific Ring of Fire' and built on unstable soft soil.

Los Angeles (USA)	15	2000	Potential of 50 000 injured and $200 billion worth of damage. Situated on the Cascades Subduction Zone. A 99% probability of an earthquake in the foreseeable future. Would have a major international effect because of the economic importance of the city.
Istanbul (Turkey)	14	55 000	Junction of the African and European Plates. A study in 2000 found the city faces a greater than 60% chance of experiencing a magnitude 7.0 earthquake by 2030.

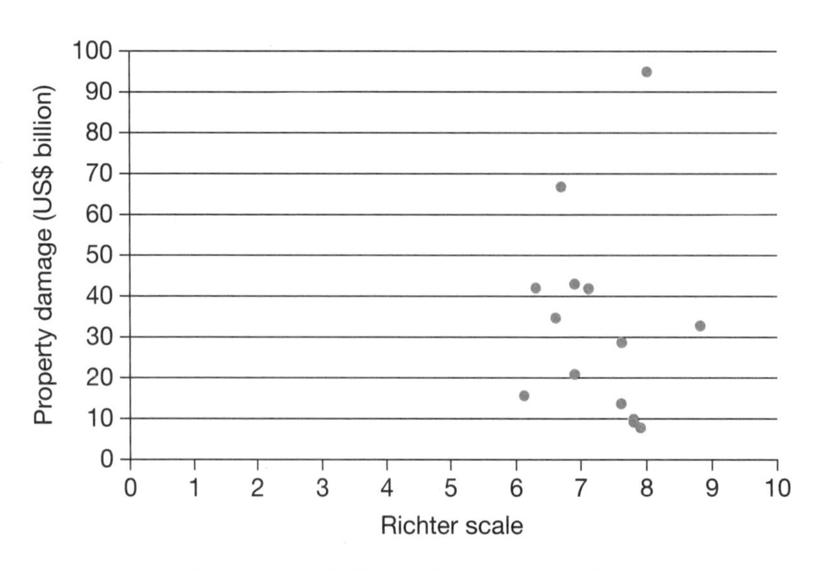

Some twentieth century earthquakes

Figure 2

Turkey's earthquake risk

Turkey is no stranger to deadly earthquakes. More than 100000 people died across the country during earthquakes in the twentieth century, according to the World Bank, and that risk hasn't abated in recent years.

Turkey has invested in retrofitting public buildings in Istanbul, its most populous city, but most of the city's inhabitants live in hastily constructed homes that do not meet buildings codes.

A magnitude 7.6 earthquake near Ízmit in the Kocaeli province of Turkey in 1999, 75 miles away from Istanbul, killed nearly 1000 people in Istanbul alone. A similar earthquake in the city centre would kill many more.

	Location	Year	Richter scale	Fatalities
1	Erzincan	1939	7.8	32700
2	Ísmit	1999	7.6	17118
3	Ladik	1943	7.6	4000
4	Malazgirt	1903	7.0	3500
5	Mürefte	1912	7.8	2800
6	Gerede	1944	7.4	2790
7	Varto	1966	6.8	2529
8	Lice	1975	6.7	2000
9	Erzurum	1983	6.9	1342
10	Ustukran	1946	5.9	1300

Source: USGS

Major devastating earthquakes in the twentieth century, by number of fatalities

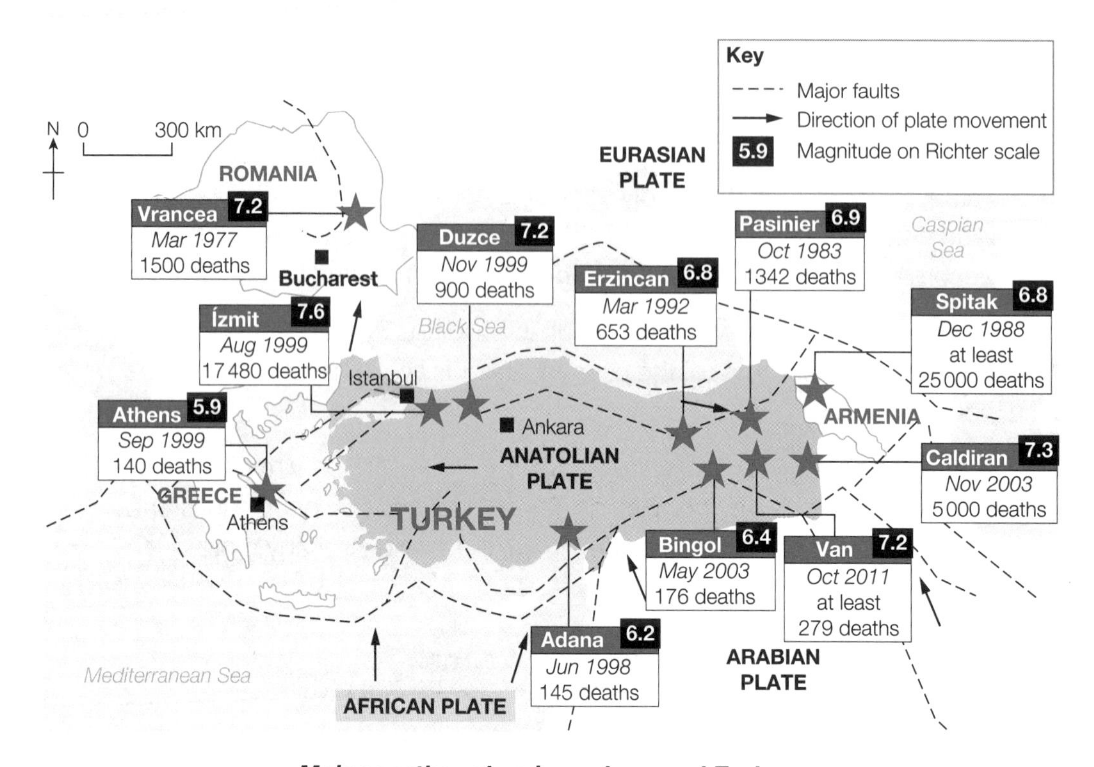

Major earthquakes in and around Turkey

Turkey is particularly vulnerable to earthquakes because it sits on major geological fault lines.

- Two earthquakes in 1999 each with a magnitude of more than 7 killed almost 20 000 people in densely populated parts of the north-west of the country.
- Two people were killed and 79 injured in May 2017 when an earthquake shook Simav in north-west Turkey.

Van earthquake 2011

Turkish earthquake in Van province kills 272 and injures more than 1300

An earthquake measuring 7.2 on the Richter scale devastated parts of Van province in the east of Turkey. Hundreds are still missing, trapped beneath the collapsed ruins of buildings with frantic survivors digging into the rubble with their bare hands in a desperate attempt to rescue the trapped and injured.

Hospitals are amongst those buildings that have been seriously damaged, meaning emergency medics have had to treat casualties outdoors. In the immediate aftermath, many of the bodies of those that had died were left next to the ruined hospital buildings. The Foreign Secretary William Hague has offered aid to the stricken area.

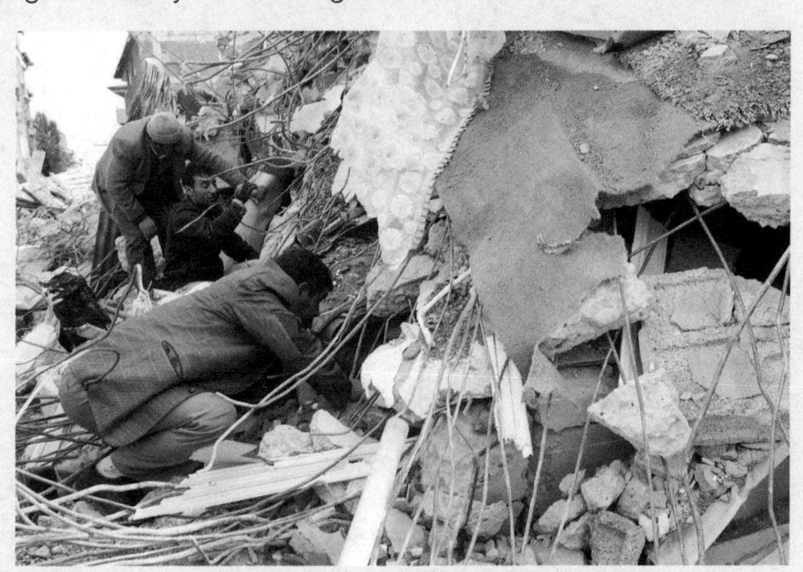

Rescuers searching for survivors

The death toll has been confirmed at 272 with more than 1300 injured, although these figures are likely to rise due to magnitude of the earthquake and the poor standard of housing.

A spokesman commented: 'As the rescue work progresses, it is likely that the death toll will increase but we are talking about hundreds still trapped rather than thousands.' Tens of thousands have been made homeless and have been sleeping outdoors through the night in freezing conditions.

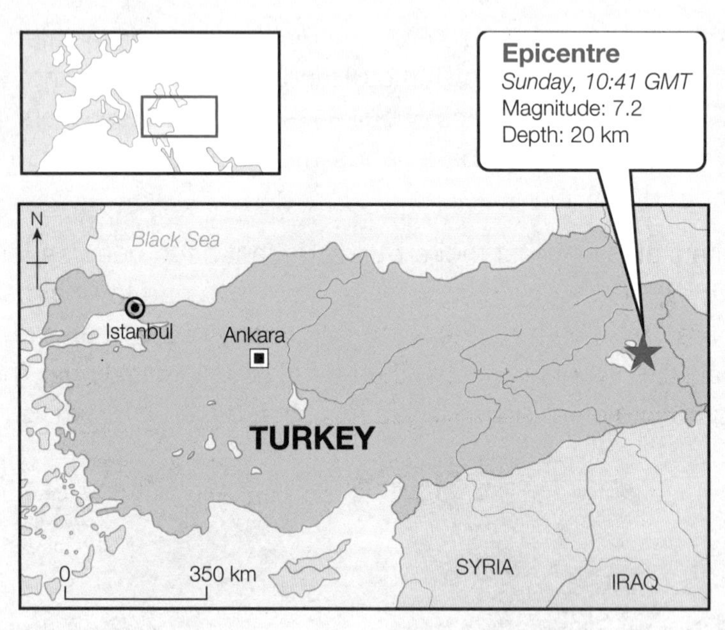

Location of Van earthquake, 2011

The relief effort numbers 2400 rescue workers, 680 medics, 12 rescue dogs and 108 ambulances, including seven air ambulances.

The region has been hit by more than 200 aftershocks, hampering rescue work and preventing people returning to their homes.

Survivors have been sheltered in camps and supplied with blankets, food and water that have been supplied along with mobile kitchens. The military have supplied aircraft in helping with the rescue and relief efforts.

Emergency teams continued into the night to try to rescue people believed to be trapped in the ruins of buildings. Van city, near the Iranian border, is a bustling city with many apartment buildings.

The situation appeared chaotic as night fell. There were too few officials to organise the rescue efforts and to give help to the shocked and bloodied survivors, as aftershocks continued to rock the city.

Figure 3

Istanbul

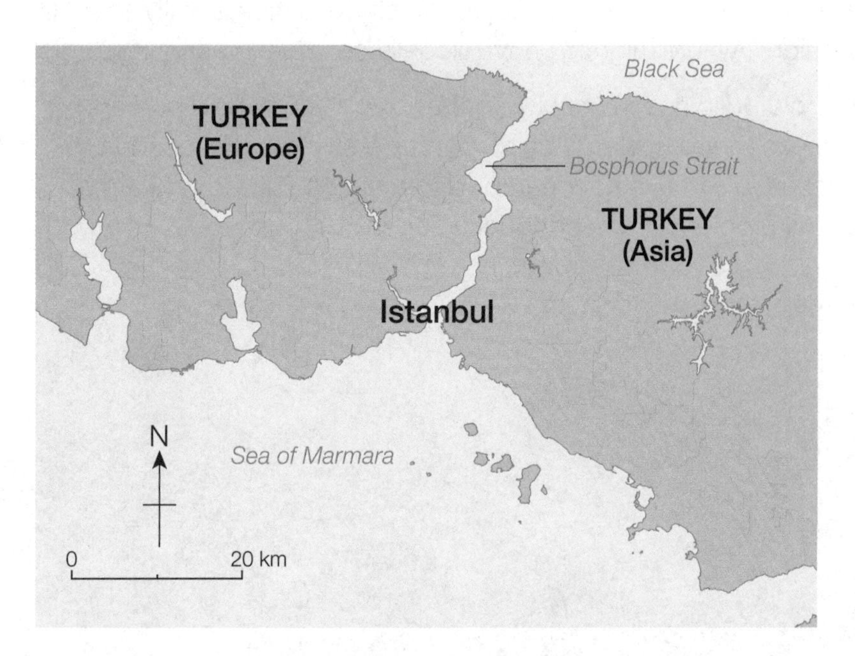

Location of Istanbul, Turkey

Istanbul has about 20 per cent of Turkey's population yet produces nearly half of the country's wealth. It is an important textile manufacturing and food processing centre but the fastest-growing sectors are finance and tourism.

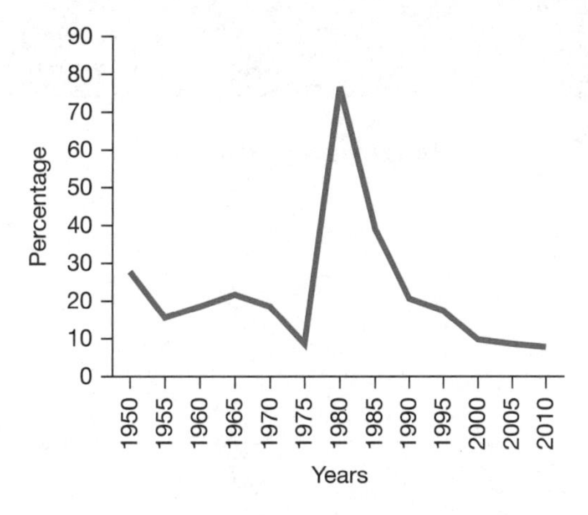

Istanbul's growth in five-year intervals

In 1960, 20 per cent of Turkey's population was urban. This figure had risen to 70 per cent by 2015 due to rural–urban migration, with Istanbul showing the greatest increase. This has led to the growth of squatter settlements called 'gecekondular', which translates literally as 'built overnight', where migrants have illegally occupied vacant land.

Also, old buildings in the city centre have been subdivided and have expanded upwards by adding new stories. These, and the gecekondular, have not been built to a high standard and are a serious problem in an earthquake zone.

Aerial view of Istanbul

The Marmara fault

The red line indicates the offshore Marmara fault where a major earthquake is overdue.

The green rectangle shows an area where there have been a number of earthquakes – there is movement along the fault in this area.

The area of the fault within the blue rectangle near Istanbul is locked, so pressure is building up in this area.

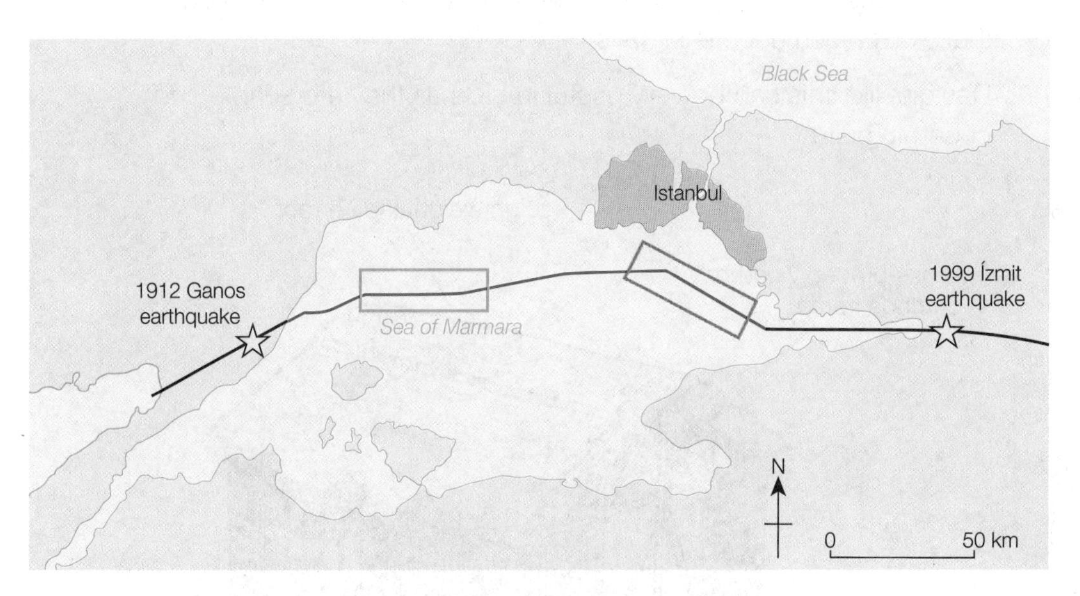

Location of the Marmara fault

Saving lives in earthquake zones

Building to survive

It is sometimes said that earthquakes don't kill people, but buildings do. But if buildings are carefully designed the people inside will survive an earthquake even if the surrounding buildings fall down around them. Even simple buildings can be made safer.

The design of the building in the diagram below has secure foundations to support and hold the building together and a light roof that won't crush people if it falls.

Designs like this can be very useful in LICs as they are simple and cheap to build.

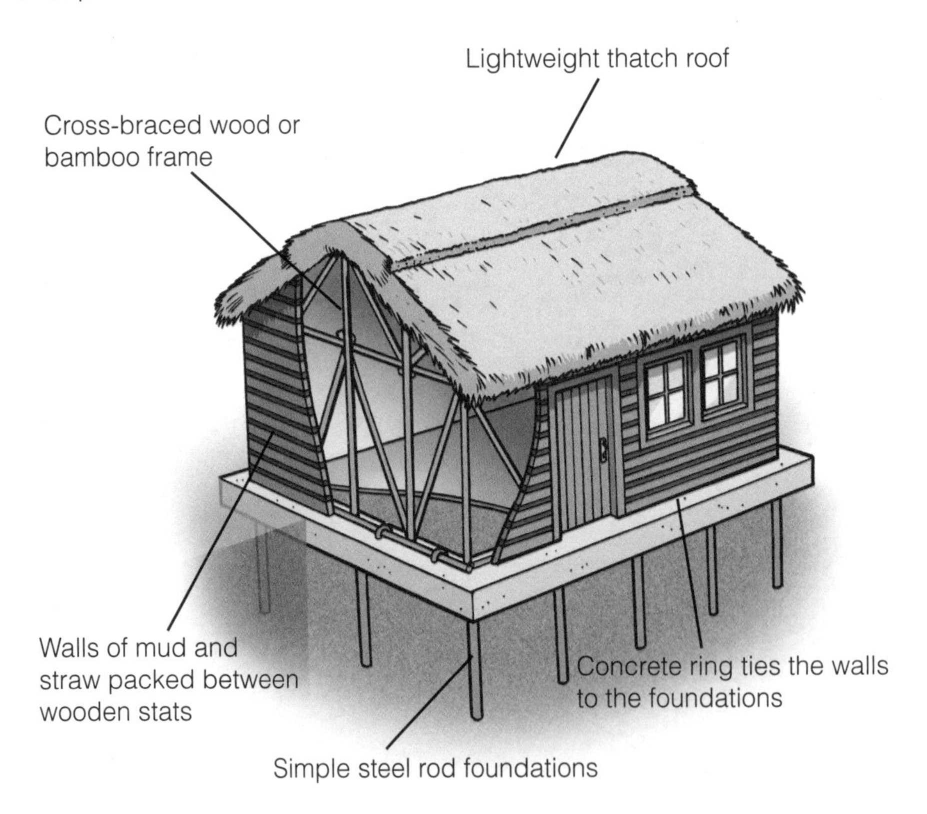

Simple building design can make a big difference

HICs have more money to spend on designing buildings to withstand earthquakes. A modern 'earthquake-proof' building uses strong building materials, shock absorbers and a damper in the roof to minimize shaking and damage (see the diagram below).

These buildings are designed to reduce the damage likely to be caused by an earthquake. This is known as mitigation.

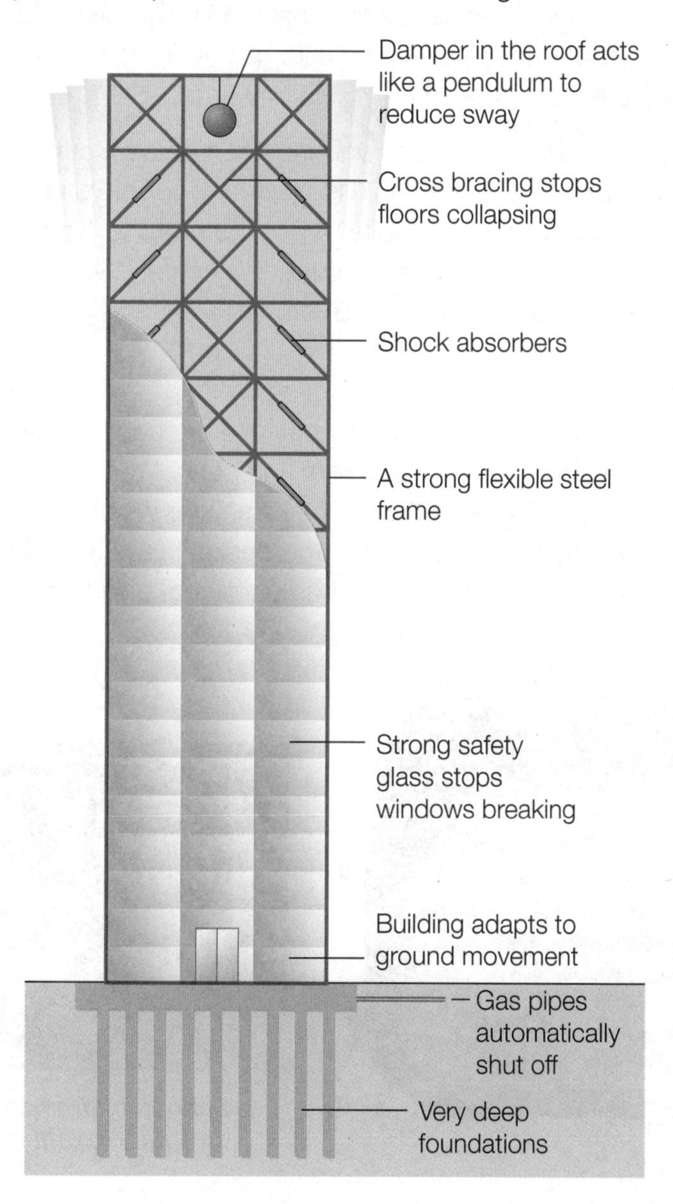

Building to survive future quakes

Prediction and warning systems

There is no way to predict earthquakes accurately. However, data recording the movement of plates are collected from satellites and seismometers. This information can be used in GIS to produce hazard maps so that warnings can be given.

One way to identify areas that could experience earthquakes is to look for seismic gaps. These are areas along known faults that haven't experienced earthquake activity for a long time. These areas are likely to be where pressure is building.

Once seismic waves start, more warnings can be sent out. When the Tōhuku earthquake hit Japan in 2011, residents in Tokyo received a 30-second warning on their mobile phones. Smartphones are also being developed to measure the ground shaking and warn people when an earthquake is about to happen.

Be Prepared

If people in areas prone to earthquakes have enough money, they can make sure they are prepared. Workers and school children in Japan do regular drills. This means they will take shelter when the warning is given. In school this would be under the desks. Ideas such as a 'go bag', can also help people survive if an earthquake strikes.

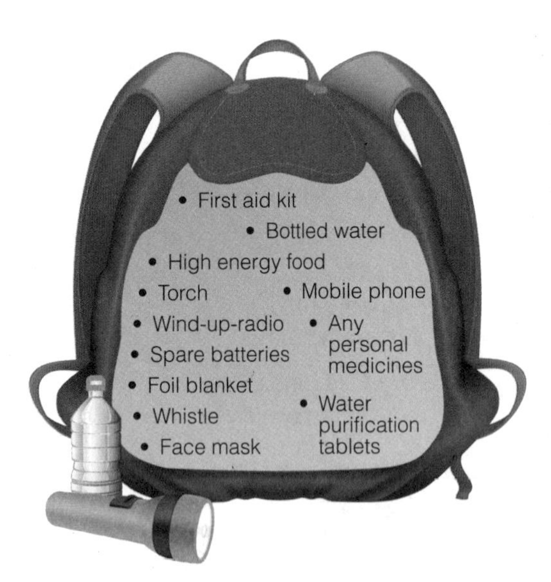

A 'go bag' contains a range of survival equipment

Schoolchildren carrying out earthquake drill in Japan